THE PACHAMAMA HYPOTHESIS

A Radical Hypothesis that Life Alters its Environment for its Benefit and Created the High Biodiversity of the Earth

by David Seaborg

Komodo Dragon Publishing
Walnut Creek, CA
2025

Cover photo credits:

Tiger by George Desipris

Sphinx Moth by Roger Lee

Sea Jellies by Michal Pechardo

ISBN: 978-0-9969513-9-5

DEDICATION

I dedicate this book to:

Lynne Cobb, Lela Cobb, Molly Cobb-Holderness, Adele Seaborg, Eric Seaborg, Steve Seaborg, Sumer Barcelon, David Hofstetter, Roger Jakobson, Megan Kalyankar, Alex Loijos, Dallas Martin, Ruth Sheridan, Benjamin Tubbs, Hovig Bayadorian, Steve Garan, Mark Koperweis, Sonoko Masui Rooney, Hal Segelstaad

ACKNOWLEDGEMENTS

Sterling Bunnell encouraged and inspired me and had several valuable discussions with me about the Autocatalytic Biodiversity Hypothesis/Pachamama Hypothesis and my ideas before I began writing this book or gave the theory its name. He had similar ideas to mine.

The following scientists discussed the Autocatalytic Biodiversity Hypothesis/Pachamama Hypothesis with me, encouraged me, and gave me confidence and inspiration, the latter three before I began writing the book or gave the theory its name:

James Lovelock, Lynn Margulis, Eugene Odum, and Edward O. Wilson.

The following scientists helped me with ensuring that the science and logic are correct:

Frank Almeda, Roni Avissar, Robert Burner, Donald E. Canfield, David Catling, John Coffin, Romain Darnajoux, Terry Erwin, Paul G. Falkowski, Sergey Gavrilets, Scott F. Gilbert, Robert Hazen, James Kasting, Joseph Kirshvink, Eugene V. Koonin, Lee R. Kump, Mark Leckie, Egbert Giles Leigh, Jr., Timothy M. Lenton, Aditee Mitra, Clenton Owensby, Mary Power, Jaime Iranzo Sanz, Alex L. Sessions, Geerjat Vermeij, Luis Villarreal, Yuk L. Yung, and Frank Zindler. Frank Almeda and Aditee Mitra were especially helpful in these discussions.

Sylka Perez, Roni Avissar's assistant, was very helpful in conveying my messages.

Of those above, the following scientists gave extraordinary help and a great deal of their time, discussing ideas and/or reading, commenting on, and/or suggesting changes to parts of the manuscript, and giving invaluable feedback to assure scientific accuracy:

Robert Burner, John Coffin, Romain Darnajoux, Jaime Iranzo Sanz, Eugene V. Koonin, Clenton Owensby, Luis Villarreal, Yuk L. Yung, and Frank Zindler.

The following laypeople read chapters of the manuscript, doing valuable editing, improving the English and presentation of my ideas, including making the wording more succinct, comprehensible, interesting, and exciting:

Hovig Bayandorian, Lynne Cobb, Steve Seaborg, and Reid Stuart.

The following laypeople sent me articles supporting the hypothesis that were helpful because they addressed aspects that I was at least partially unaware of:

Lynne Cobb, Bob Jansen, Adele Seaborg, and Steve Seaborg.

Lynne Cobb sent me the book, *The Medea Hypothesis*, by Peter Ward, making me aware of a hypothesis that is in opposition to the Autocatalytic Biodiversity Hypothesis/Pachamama Hypothesis.

Mamade Kadreebux encouraged me constantly and provided me a forum from which to speak to an audience about the hypothesis.

Adele Seaborg solved computer issues that were beyond my limited knowledge in that area.

Reid Stuart did page layout.

Countless teachers, extended family members, friends, and scientists I knew, but did not speak with specifically about this book, made this book better. So did many scientists I never met, but whose works I am familiar with, and gurus and teachers of philosophical systems I met or only know through their teachings. So, too, did innumerable living organisms, especially snakes; museums; and beautiful ecosystems in nature. It would not be in keeping with the spirit of this book to fail to acknowledge these.

Any errors of any sort in this book are solely my fault and not the fault of my commensal helpers.

TABLE OF CONTENTS

LIST OF ILLUSTRATIONS

CHAPTER 1: A NEW HYPOTHESIS THAT LIFE GENERATES HIGH BIODIVERSITY

What would the world be, once bereft
Of wet and of wildness? Let them be left,
O let them be left, wildness and wet;
Long live the weeds and the wilderness yet.

From the poem, "Inversnaid" by Gerard Manley Hopkins

This book is about a novel idea in evolution and ecology that proposes that the high biodiversity of the Earth was created primarily by biological organisms molding the biosphere into environments favorable to life.

There is a glossary at the end of this book, which I encourage readers who do not understand some of the terms used to consult. This paragraph defines a few words that will be helpful to many readers; those familiar with basic cell biology and taxonomy can skip this paragraph. *Eukaryotes* are organisms whose cells have most of their DNA contained in a *nucleus* within each of their cells, and whose cells contain *organelles*, which are structures in a cell that perform one or more specific functions. Examples of organelles are the *mitochondrion* (plural *mitochondria*), which carries out cellular respiration, using oxygen to produce energy for the cell, and the *chloroplast*, which carries out *photosynthesis*, which uses the energy of sunlight to produce oxygen and carbohydrate (food) for the cell and organism. The nucleus and organelles are surrounded by membranes. Eukaryotes include everything from one-celled amoebas to insects to humans. Although most of the eukaryote's DNA is in its nucleus, mitochondria and chloroplasts have their own genomes with their own DNA. The *genome* is the complete set of genes or genetic material present in a cell or organism. In contrast to eukaryotes, *prokaryotes* are simpler organisms that do not have a nucleus, but have their DNA scattered throughout their cell, and have no organelles or internal membranes. Almost all prokaryotes are organisms with just one cell; they are microbes. They are less complex than eukaryotes and include *bacteria*, which constitute one of the two major groups of prokaryotes. The other group is the *archaea* (singular is *archaeon*), which sometimes live in

1

extreme conditions such as in hot springs that have very high temperatures. Organisms are classified into categories, which are, from the largest to the smallest: *domain, kingdom, phylum* (plural, *phyla*), *class, order, family, genus* (plural, *genera*), and *species*. The domain contains more than one kingdom, which contains more than one phylum, and so on. There are only three domains of life, the largest category. They are the eukaryotes, bacteria, and archaea, all defined above. Bacteria and archaea differ from each other enough to be in separate domains, meaning they are very different from each other indeed. *Unicellular organisms* are organisms made up of only one cell. *Multicellular organisms* are organisms with more than one cell, and may be very large; they include trees, insects, whales, and humans.

Ecology, population biology, and evolutionary theory have served us well, but have given far too little emphasis to the role that organisms play in increasing biodiversity. Darwin and evolutionary theory correct state that the environment affects organisms (via natural selection), including their structure, behavior, and genetics, and it played an important role in creating today's great diversity of life. Yet it works both ways, because organisms affect the environment, and not randomly. They affect it in a way that aids life and increases biodiversity. There is feedback between life and the environment, each affecting the other.

This book is about my hypothesis that I call the Autocatalytic Biodiversity Hypothesis (ABH) because it proposes that life causes biodiversity to increase; hence, life promotes life and its own diversification, so is autocatalytic. I also call it the Pachamama Hypothesis (PH), following the naming of hypotheses that address the effects of organisms on biodiversity after goddesses. For example, the Gaia Hypothesis, proposed by James Lovelock (1972, 1979) and Lovelock and Lynn Margolis (1974), postulates that organisms were instrumental in helping make the atmosphere (and hence the climate) favorable to life. Gaia is one of the Greek primordial goddesses, the personification of the Earth and ancestral mother of all life. The Pachamama Hypothesis was influenced by and is similar to the Gaia Hypothesis, but much more general. The Medea Hypothesis, proposed by Peter Ward (2009), states that life is suicidal and decreases diversity or biomass. This hypothesis is counter to and in opposition to both the Gaia Hypothesis and the ABH/PH, proposing essentially the opposite of the latter two hypotheses. In Greek mythology, Medea is the destructive granddaughter of the sun god Helios. She killed her children according to some versions of her story. Pachamama is the ever-present Earth mother goddess who has her own self-sufficient and creative power to sustain life on Earth. She is revered by indigenous peoples of the Andes, such as the Quechua and Aymara. Like Gaia, Pachamama is an Earth goddess. Thus, the name of the Pachamama Hypothesis follows the precedent of naming hypotheses that concern life's effects on our planet after Earth goddesses or mythological figures.

I originally proposed the ABH/PH in two books that describe it and present evidence for it (Seaborg, 2021 and 2023). I have revised my hypothesis since these books were published. The following paragraphs present its current form.

The ABH proposes that *life increases biodiversity* by the following four mechanisms:
(1) *Natural ecosystems maximize biodiversity.* Since biodiversity is a form of information, one can state this by saying *natural ecosystems maximize their information content.* Biodiversity is generally the number of species, although it is more complicated than this; its definition is discussed in more detail in Chapter 2. Natural ecosystems increase biodiversity by mechanisms 2 and 3, which follow.
(2) *The net impact of every species is to make its ecosystem better for life and other species, and cause a net increase in biodiversity, in a natural ecosystem over a time period sufficient for a fair assessment of its effects.* This principle states that every species has a *net* positive impact on its ecosystem and other species, and causes a *net* increase in biodiversity because species can have some negative effects on their ecosystem, other species, and biodiversity. For example, species that compete with each other will negatively impact each other. The ABH states that the net effect of each species on its ecosystem, other species, and biodiversity is positive when all impacts of the species in question are added together. A species may have positive impacts by regulating the population of its prey, providing food to its predators, being symbiotic with other species, providing a home to other species, and other mechanisms. The positive effects that each species has on diversity

and its ecosystem outweigh the negative ones. A *natural ecosystem* is one in which human effects are not significant. Human alterations of natural ecosystems and the climate, and the introduction of exotic species are not natural situations. Humans are therefore not included in the ABH. They are an exception. Chytrid fungi have caused the extinction and reduction of several amphibian species only because of the unnatural situations of human-caused climate change and their being introduced to environments that they are not native to by humans. Herbivores denude areas of vegetation and reduce diversity greatly only when humans remove their natural predators. And modern, technology-based humans are an exception to the ABH; they make their environment worse for life and decrease diversity. *Sufficient time for a fair assessment of a species effects on its ecosystem* needs to be added because sometimes species decrease environmental quality and biodiversity in the short term, but have a positive effect on these qualities in the long run. For example, even in natural ecosystems, migratory locusts can greatly reduce vegetation and perhaps even plant diversity for a period of time, but they eventually have their numbers reduced or move to another area, and vegetation recovers. They serve as a food source for many species and fertilize the soil when they die, with the result that their net impact is positive and increases diversity, over sufficient time in natural conditions. The advent of oxygen-producing photosynthesis caused a great increase oxygen in the atmosphere and the sea. Oxygen was toxic to life at the time, and it also combined with and depleted the greenhouse gas methane, causing Earth's temperature to plummet. These two effects both negatively impacted biodiversity. But after a long time period, organisms evolved the ability to use oxygen as an energy source, allowing higher life to evolve and diversify tremendously. Over the long term, the net effect of the first oxygen-producing photosynthetic organisms, which were a form of bacteria called cyanobacteria, was to greatly increase biodiversity. Organisms increase the diversity of their ecosystem in three ways, so this mechanism 2 has three mechanisms, none of which are original ideas on their own. (a) *Some species alter their environment* by such means as predators and parasites regulating the populations of their prey and hosts, plants holding the soil in place with their roots, bees pollinating flowers, and so on. (b) *Some species provide a resource to other species simply by existing.* Prey species provide food for their predators, many species provide homes to other species, trees and coral provide three-dimensional habitat to several species, and so on. (c) *Individual organisms in some populations often cooperate with and act altruistically toward each other, maintaining the population or species, its variability and biodiversity, and sometimes increasing biodiversity.* This helps the population or species survive, maintaining biodiversity. Altruistic sacrifice for other organisms appears at first glance to contradict natural selection, but it does not. It can be accounted for by kin selection, where individuals pass on their genes by sacrificing for close relatives, which indirectly pass on their genes to the next generation, as worker bees do for the queen bee; reciprocal altruism, where individuals help each other, forming alliances and friendships; and group selection, where groups with cooperating, altruistic, helpful individuals out-complete groups of selfish individuals of the same species. Further discussion of this topic is beyond the scope of this book. It is included here for completeness.

(3) *Organisms profoundly changed Earth, including its atmosphere, land, and bodies of water, making it highly favorable to life.* They brought the level of oxygen in the atmosphere from less than 1% to a life-favorable 20%. Organisms have regulated Earth's temperature, keeping it favorable to life. The sun has been getting hotter since life began. When life had just started and the sun radiated much less heat, organisms produced the powerful greenhouse gas methane, which heated the Earth. As the sun grew hotter, organisms buried carbon, which reduced greenhouse gases and cooled the Earth. Biology is the principal force that built the soil. Organisms may have regulated the sea's salt content, resulting in levels of salt in the ocean favorable to marine life. Organisms significantly modified the Earth in other ways that helped life flourish, evolve, and diversify.

(4) *In nature, where humans are not manipulating genomes, the behavior of all genomes results in genetic variability, and sometimes evolutionary innovations; the latter leads to diversification. All genomes are creative, not consciously, but as a result of physics, chemistry, and electromagnetic interactions. Their actions often result in increases in biodiversity.* The Pachamama Hypothesis can be summarized as follows: *organisms create biodiversity; life makes biodiversity significantly higher than would be the case if solely non-biological factors such as physical, chemical, and geological forces acted in the absence of life; and ecosystems maximize biodiversity.* A simple way to express this hypothesis is that *life creates more species.* This does not fully

5

capture the meaning of the idea because biodiversity is more than simply the number of species. Science seeks unifying principles, and *the Pachamama Hypothesis is a unifying principle* that brings together how plants, herbivores, predators, decomposers, bacteria, viruses, and other life forms all promote biodiversity. To state this unifying principle simply: *in a natural ecosystem, the net effect of each species over time is to maintain and increase the diversity of its ecosystem.* If one desires to state the ABH in a four-word phrase, it is: *life creates more species.* It is well known that symbiotic and commensal species help other species, but what makes the ABH *original* is that it states that *every* species has a net positive effect on biodiversity, other species, and its ecosystem. Species do this by the mechanisms discussed in this book: providing food, providing habitat, regulating prey populations, predators selectively attacking the superior competitor, regulating temperature and greenhouse gasses, increasing and regulating oxygen, making nitrogen available to life, enhancing habitat, donating genes to other species, and other mechanisms. This is supported by the observation that the removal of any species from an ecosystem generally reduces the number of other species, or at least reduces the populations of some species.

Every species in a natural system increases its ecosystem's biodiversity and has a positive impact on its ecosystem because each species is regulated by other species: by its predators, parasites, and competitors. If a species were allowed to increase its population indefinitely, it would have a net negative impact on its ecosystem, other species, and biodiversity. The field mouse (*Microtus pennsylvanicus*) benefits its ecosystem by making tunnels that aerate the soil; regulating some species of of grasses, sedges, and forbs; fertilizing the soil; providing habitat to an ecosystem of microbes in and on it; and providing food to its predators. This is true because its predators and parasites regulate its population. If they did not do so, its population would grow to the point where it depleted its supply of grasses, sedges, and forbs, greatly harming its ecosystem and decreasing its biodiversity. This is true of all species of organism. *It is the entire interacting ecosystem that ensures that every species helps its ecosystem and increases biodiversity.* This is another reason why the ABH states that ecosystems maximize biodiversity.

The other mechanism by which ecosystems maximize biodiversity is by coevo-

lution. All strongly interacting species coevolve with each other. This allows the evolution of symbiosis. It allows predators and prey, and parasites and hosts, to be in balance. Coevolution allows competing species to evolve to differ so much that they no longer compete with each other. It is the *coevolution* of the field mouse with other species that causes it to have a net positive effect on its ecosystem, other species, and biodiversity. *Biodiverse ecosystems are stable because the species in them coevolve.* An ecosystem with a million species of predators would have high diversity, but would not be stable. A stable ecosystem has a good balance of plants, herbivores, predators, and decomposers. This comes about by coevolution. Thus, biodiversity alone does not cause the stability of an ecosystem. *Coevolved biodiversity causes an ecosystem's stability.*

Since the ABH/PH states that every species increases the biodiversity of its ecosystem, the number of species should increase indefinitely through time. Yet it seems unlikely that this process can go on indefinitely, with species number increasing infinitely. Limited availability of resources, such as light, water, nutrients, and space should limit species number. Whether a plot of species number against time results in a logistic curve where a carrying capacity is reached, the curve is asymptotic, or takes on another form, is not known and can only be determined by further research.

Symbiosis is a relationship between two species in which each benefits the other. *Commensalism* is a relationship between two species in which one benefits and the other is not affected. They will be defined in more detail in the chapters about them. *Related to the ABH is the idea that symbiosis and commensalism are very common, pervasive, fundamental, important, help structure ecosystems, affect evolution, and are very important in maintaining and increasing biodiversity. They are far more important than commonly recognized. Both of these relationships between species are tied to the ABH because the principal mechanism by which the ABH works is that all species have a net positive effect on biodiversity and their ecosystem.*

Another idea related to the ABH is that all genes are connected to other genes, the genetic system of the organism they are in, the gene pool of the population they are in, and the genes of the ecosystem they dwell in. No gene exists or evolves autonomously. Genes coevolve with each other. Organisms are systems of coadapted genes. Similarly, all organisms of a population are inter-related and no organism exists or evolves in isolation. For example, animals, plants, and fungi are connected to the communities of microbes that live within and on them, other members of their species, and other organisms in their community. Likewise, *all species are connected and no species exists or evolves in isolation. Each species is connected to other species in its com-munity and to its ecosystem. Coevolution between species is fundamental and common. All species that strongly interact coevolve with each other. Finally, ecosystems are connected to and interact with other.*

The ABH was influenced by the Gaia Hypothesis and I am indebted to Lovelock and Margulis for this. The Gaia Hypothesis and ABH have areas of overlap, but they are not identical. First, the ABH is much more comprehensive and general than the Gaia Hypothesis, which is a subset of the ABH. The main point that the Gaia Hypothesis postulates is that life created an atmosphere favorable to life. The ABH also proposes this, but it also includes mechanisms of increasing biodiversity via natural selection, symbiosis, commensalism, plants, herbivores, predators, decomposers, bac-teria, viruses, soil organisms, as well as other mechanisms.

Second, the Gaia Hypothesis proposes that life helps life by negative feed-back on the atmosphere. The ABH states negative feedback has a role, in-cluding life acting on the atmosphere, but this can be over-ridden, and negative feedback acts most effectively in life-on-life systems like predators and prey. Moreover, positive feedback can increase diversity in such situa-tions as symbiotic coevolution.

Finally, the Gaia Hypothesis does not propose that any particular variable in an ecosystem is maximized by life. The ABH postulates that life's activi-ties maximize an important variable of the ecosystem: biodiversity. Specifi-cally, the ABH postulates that ecosystems maximize biodiversity.

Many supporting concepts, observations, and facts presented here are not original in themselves. *What is original about the ABH is its integration of the facts and evidence into the comprehensive hypothesis that life gener-ates biodiversity.* Hence, when, for example, I provide an explanation of the microbiome—that animals and plants are ecosystems that have tens of thousands of species of microbes on and in them—I do not claim that this is an original idea. Rather, it is a piece of evidence showing that organisms create and maintain biodiversity. To my knowledge, this work is the first attempt at a comprehensive compilation of evidence that life itself creates biodiversity.

The ABH is an original, radical, testable, qualitative (not mathematically expressed as yet) hypothesis supported by the evidence. It is scientific, mechanistic, and naturalistic. It is not teleological or "New Age", and so does not posit a design or purpose, or a conscious Earth or outside consciousness, directing it.

References

Lovelock, J. E. (1972). Gaia as seen through the atmosphere. *Atmospheric Environment* 6 (8): 579–80. Bibcode: 1972AtmEn...6..579L. doi: 10.1016/0004-6981(72)90076-5

Lovelock, J. E. (1979). *Gaia. A New Look at Life on Earth*. Oxford Univ. Press, Oxford, London.

Lovelock, J. E. & Margulis, L. (1974). Atmospheric homeostasis by and for the biosphere: the Gaia hypothesis. *Tellus*. Series A. Stockholm: Internat. Meteorol. Inst. 26 (1–2): 2–10. Bibcode: 1974Tell...26....2L. doi: 10.1111/j.2153-3490.1974.tb01946.x. ISSN 1600-0870

Ward, P. (2009). *The Medea Hypothesis: Is Life on Earth Ultimately Self-Destructive?* Princeton Univ. Press, Princeton, NJ, and Oxford, UK.

Seaborg, D. (25 Sept. 2021). *How Life Increases Biodiversity: An Autocatalytic Hypothesis*. 264 pp. CRC Press; Taylor & Francis Group. Boca Raton, FL; London, U.K.; New York, NY. ISBN 9780367631345. https://doi.org/10.1201/9780429440137. eBook published 8 Sept. 2021. eBook ISBN 9780429440137.

Seaborg, D. (2023). *Organisms Amplify Diversity: An Autocatalytic Hypothesis*. 254 pp. CRC Press; Taylor & Francis Group. Boca Raton, FL; London, U.K.; New York, NY. ISBN vv9781032158020. https://doi.org/10.1201/9781003246640. eBook published 29 June 2023. eBook ISBN 9781003246640.

CHAPTER 2: BIODIVERSITY: ORGANISMS CREATE IT, ECOSYSTEMS MAXIMIZE IT

What Is Biodiversity?

Because my thesis claims that organisms increase biodiversity, it is important to precisely define it. It is the variability among living organisms, including diversity within species, between species, and within ecosystems. It is the information in a system. A greater variability of DNA represents more information. As a first approximation, biodiversity is species richness, which is defined as the number of species in a community. Biodiversity, is also used to mean species diversity, which is both species richness and the evenness of distribution of organisms within different species in a community. Thus, a system with ten species with 100 organisms in each species has higher species diversity than one with ten species and 910 organisms in one species and ten organisms in each of the other nine species, even though each has the same number of species and organisms. But biodiversity is more than species diversity. It also includes variation within species and populations. This includes genetic variability between individuals in the same population or species, as well as between different populations, races, and subspecies of the same species. Another measure is the number of species in a higher taxon, such as within a genus or family; number of genera within a family; families within a class; or any lower taxonomic level within a higher one. The term can also be used to refer to the relative number of species or other measures of diversity within different ecosystems. Biodiversity can be defined as the variation of life forms within a given ecosystem, within a given biome, or on the entire Earth.

Numbers of Organisms and Biodiversity Were Both Spectacularly High Before Human Impacts Became Dominant

Both population sizes and numbers of species were extremely high before the Industrial Revolution. People reported salmon so dense that one could not see the stream bottom and could walk on them. There were whales so dense that the people on the beach in the 1800s complained of the stench of whale breath. Dasmann (1965) documented the amazing abundance of Californian animals and plants. California poppies once covered hillsides in continuous orange. The journalist MacKinnon (2013) cited records from recent centuries that show the abundance of wildlife in those times: "In the North Atlantic, a school of cod stalls a tall ship in mid-ocean; off Sydney, Australia, a ship's captain sails from noon until sunset through pods of sperm whales as far as the eye can see... Pacific pioneers complain to the authorities that splashing salmon threaten to swamp their canoes."

Global Species Number is Tremendous

There were flocks of birds that took days to fly by in the sky, blackening the skies. Even the now-extinct passenger pigeons darkened the sky with their numbers for long periods. There were lions in the south of France, walruses at the mouth of Britain's Thames River, and 100 blue whales in the Southern Ocean for every one that is there today. The world's largest king penguin colony shrank by 88% in 35 years, and more than 97% of the bluefin tuna that once lived in the ocean are gone, meaning these two species were recently many times more abundant than they are today.

Erwin (1982), measured insect diversity in a Panamanian tropical rainforest canopy, concluding that there could be as many as 30 million tropical, terrestrial arthropod species extant globally, far exceeding the usual estimate of 1.5 million at the time of the study (ibid). This extrapolates to about 47 million arthropod species worldwide, yielding an estimated total of approximately 56 million animal species on Earth. If human effects lowered the species count in Erwin's study, the estimate for total species on the planet before human impacts would be even higher than 56 million. There may be many more than this because 30% of all species could be cryptic species. Mora et al. (2011) did a study that suggests that 86% of the existing species on land and 91% of species in the sea are as yet undescribed. Quentin Wheeler, an entomologist who directs the International Institute for Species Exploration said: "Our best guess is that all species discovered since 1758 represent less than

20% of the kinds of plants and animals inhabiting planet Earth". Many discovered species have not been described yet. Between 5 and 10,000 new species are found each year, most of them insects. It is estimated that only about 20% of insect species have been described; there are millions of undescribed insect species. A square foot of soil 2 inches deep is estimated to contain 22 unique mite species. A recent study estimated that there are about 18,000 bird species, twice the number previously thought to be the case (Barrowclough et al., 2016). Locey and Lennon (2016) combined a universal dominance scaling law with a lognormal model of biodiversity to predict that there are between 100 billion and upward of 1 trillion microbial species on Earth. Virus types (species) are at least ten times all cellular species combined, probably much more than this.

Life Achieved Amazingly High Diversity with Very Little Space and Nutrients

It is useful to bear in mind the small size of the biosphere, the thin layer of gases, soils, and liquids within which all life exists, to comprehend the magnificence of life's diversity. The troposphere, the atmospheric layer where life and the weather are, varies between 17 km at the equator to 7 km at the poles. The mean depth of the ocean is about 3.7 km. Thus, the entire biosphere runs from about 3.7 km below the sea's surface to about 17 km above Earth's surface, or a total of only about 20.7 km, or about 12.9 miles! Within this ultrathin film, all life's processes occur, as well as the ocean currents and the weather. The atmosphere has a total of only about 2,000 kg of oxygen and a tiny 3 kg of CO_2, of which the carbon content is about 1 kg. It is incredible that aerobic life can obtain its energy, and photosynthesizers can power life in the photic zone on this small amount of oxygen and carbon, and that life achieved such high diversity, complexity, and information content within such a small area with so little fuel.

Natural Selection Favors Biodiversity

Leigh and Vermeij (2002) pointed out that three types of evidence indicate that natural selection results in natural ecosystems being organized for high productivity and species diversity: (1) novel changes to natural ecosystems, such as human disturbances, tend to diminish their productivity and/or diversity; (2) humans must recreate the properties of natural ecosystems to enhance the productivity of artificial ones; and (3) productivity and diversity have increased after every mass extinction and during the entire history of higher life, beginning with the Cambrian explosion The Cambrian explosion is the great diversification of animals between about 541 and about 530 mya at the beginning of the Cambrian Period. Many of the major animal phyla--25 to 30 of them--that make up today's animals, appeared then. They assert that natural selection results in ecosystems organized to maintain high species diversity and productivity. Energy that is poorly exploited attracts more efficient exploiters, favoring the creation of new niches and more efficient resource recycling. Ecological monopolies that reduce productivity are eliminated. Efficiency of predators and herbivores tends to increase, favoring faster turnover of resources. All of this results in a higher number of species.

Niches Become Filled Quickly; There Are No Empty Niches for Long Time Periods

Life tends to rapidly reach its maximum possible level of biodiversity. Ecosystems always quickly go to their "carrying capacity" for their number of species. Every empty niche becomes occupied in a short time; no niche remains empty for a long period of time. The best evidence for this is that natural ecosystems that are not disturbed by humans appear to be at carrying capacity with respect to the number of species in them. Other evidence is that diversity returns to at least its previous level quickly after mass extinctions, similar niches are filled by unrelated species, and metazoan diversity has increased since the Cambrian explosion. Brayard et al. (2017) found fossils showing a rapid recovery and diversification of marine animals only about 1.3 million years after the Permian-Triassic mass extinction event, which killed off 70% of land species.

Further evidence that empty niches are quickly filled comes from the fact that if a species that normally fills a specific niche is absent, an unrelated species evolves to fill that niche. For example, the honey possum (*Tarsipes rostratus*), a tiny marsupial that feeds on the nectar and pollen of a diverse range of flowering plants in southwest Australia, and an important pollinator of many such species, occupies the niche of hummingbirds, which do not occur in Australia.

In a broad sense, deer and kangaroos occupy similar niches, both consuming leaves, shrubs, grasses, fungi, and other items. There are a number of species of each and the comparison is general, with various specific kangaroo species each having more or less the same niche as a corresponding deer species. There are several marsupials that have undergone convergent evolution with placental mammals, filling in niches that placentals would typically occupy, but do not in Australia because placentals do not live there. There are marsupial equivalents of placental moles, mice, rats, rabbits (bandicoots), raccoons (wombats), flying squirrels, ground cats, and even a groundhog. In the past, but now extinct, there were marsupial wolves, lions, and even saber-toothed cats. This is clear evidence that if a taxon is absent from a large geographic area, another taxon will evolve to fill the niches it typically exploits. This is strong evidence that empty niches do not remain unfilled for long, supporting the idea of the tendency of organisms to generate diversity and thus the ABH.

Parrott et al. (2019) found that native Australian water rats (*Hydromys chrysogaster*) have learned to flip over and eat the toxic cane toad (*Rhinella marina*), which was introduced by humans, avoiding its poison by removing the gall bladder, which contains toxic bile salts, with surgical precision, and eating the heart. They learned to avoid the toxic areas, eating only the nontoxic parts.

The cane toad is an invasive species whose populations have exploded. They have negatively impacted their prey, competitors, and predators tremendously. This shows that ecosystems in time control high-population problem species because they offer an abundant resource. The higher their numbers, the greater the amount of resource they provide, and the greater the payoff and selection pressure to adapt to exploit them. However, introduced species can greatly decrease diversity before this happens, and we should do everything possible to prevent introducing species to new areas. This also shows that when a novel niche (in this case, the toad) is added to an ecosystem from the outside, it is in time exploited by a species that fills the open niche.

Except for three bat species, New Zealand has no native mammals. So mammalian niches have been filled by insects, gastropods (the class that includes snails and slugs), reptiles, and birds. The 70 species of weta, which resemble giant crickets, eat seeds and smaller invertebrates, playing the part that mice do almost everywhere else. The South Island giant moa was a 6-foot-6-inch flightless bird, which could reach 12 feet up into trees with its neck to eat leaves, occupying a similar niche to elephants and giraffes, which it underwent convergent evolution with. The kiwi (genus *Apteryx*), a bird with five extant species, occupies a mammalian niche due to lack of mammals in New Zealand. Like mammals, but unlike ordinary birds, they have a highly developed sense of smell. They are flightless, have tiny wings, have no keel on the sternum to anchor wing muscles, have hair-like feathers, and nostrils at the end of their long beaks. Adult birds other than kiwis typically have bones with hollow insides to minimize weight for flight, but adult kiwis have bone marrow, like mammals and the young of other birds. Their sight is so poorly developed that blind specimens have been observed in nature; birds generally rely on sight more than olfaction, and mammals generally are the opposite. Kiwis prefer subtropical and temperate podocarp and beech forests, but deforestation has forced them to adapt to other niches and habitats, including subalpine scrub, tussock grassland, and the mountains, demonstrating that species can adapt to novel niches and habitats under strong selection pressure. Their nocturnal habits may be yet another niche shift from habitat intrusion by introduced predators. Where such predators have been removed, they are often seen in daytime.

Four Patterns of Biodiversity Strongly Support the ABH

Changes in biodiversity through geologic time support the ABH. The bones and shells of skeletonized marine animals have left the most complete fossil record there is. Sepkoski (1984) plotted the number of genera and families of skeletonized marine animals starting with the first complex animals, from 542 mya to present. His famous graph is shown in Figure 2.1(a), the first part of Figure 2.1. It shows a rapid increase in family number from the beginning until a plateau is reached, and then family number remains approximately constant until the first mass extinction, the end-Ordivician extinction of about 443 mya, when diversity plummets. Then diversity as measured by number of families increases until a plateau with about the same number of families as before the extinction is reached, and this remained the same until the second mass extinction. Then diversity recovered to about the same level as the first two plateaus and remained on this plateau until the third mass extinction, the Permian-Triassic mass extinction. Remember, this is the largest in Earth's history. Up to 96% of all marine and 70% of terrestrial vertebrate species went extinct. It is the only known mass extinction of insects. About 57% of all biological families and 83% of all genera went extinct. Terrestrial diversity (not marine diversity) took significantly longer to recover than after any other mass extinction event, possibly up to 10 million years. But then diversity rose *above* the old plateau, and kept rising steadily almost fourfold, until human effects became prevalent, interrupted only by the two other mass extinctions. It recovered and continued to rise after each of these two mass extinctions. Biodiversity rose at a very rapid rate and went way beyond the plateaus that existed before the Permian-Triassic mass extinction. The fossil record indicates there were just over 900 marine animal families at the last measurement point. There are about 1,900 families alive today, including those rarely or never preserved as fossils. So the family number increased greatly even since the last measurement point.

Figure 2.1(b), the second part of Figure 2.1, shows a chart by Foote (2000) of the number of genera of skeletonized marine mammals through time. It shows a similar pattern to Figure 2.1(a), with more fluctuation before the rise in biodiversity after the Permian-Triassic mass extinction. That rise gets steeper after this mass extinction. Diversity increased steadily on land after this mass extinction as well, especially after the mid-Cretaceous (Benton, 1990).

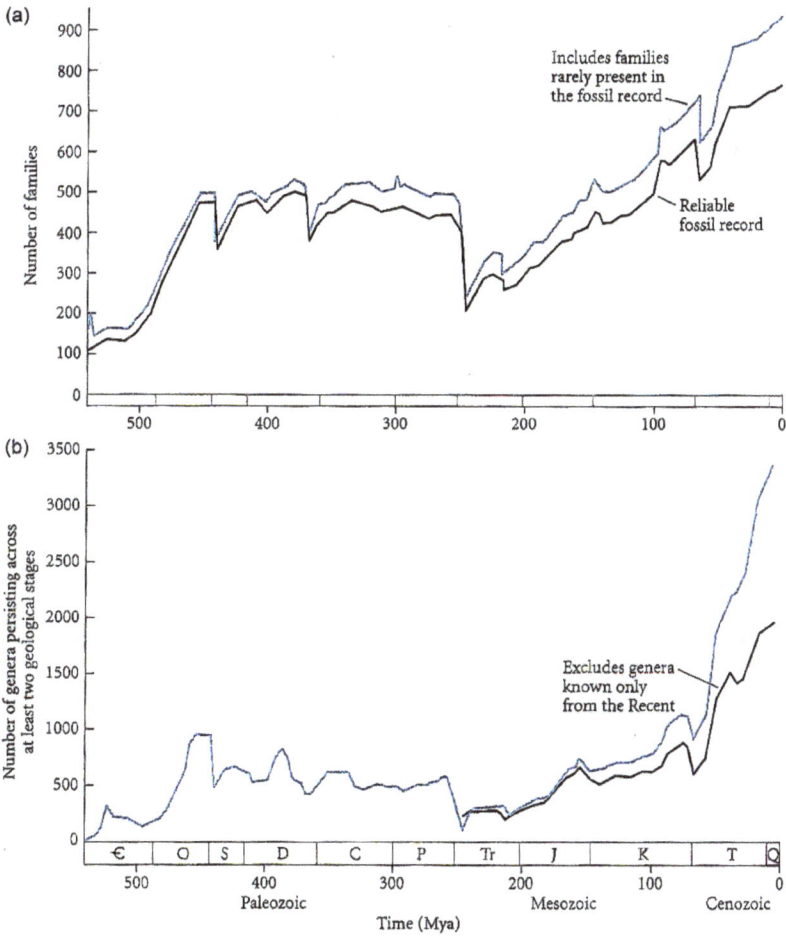

Figure 2.1. Diversity of skeletonized marine animals from the beginning of higher life and the Cambrian explosion to pre-industrial times. The number of taxa for each geological stage includes all those whose known temporal extent includes that stage. (a) Diversity of families. The black curve represents only those families with a reliable fossil record. The blue curve includes families rarely preserved. The black and blue curves concur well, so any bias from including or excluding rare forms is not important. From Sepkos-

17

ki (1984). (b) Diversity of genera of these animals, counting only those that crossed boundaries between two or more stages. The black curve excludes genera known only from the recent times, to avoid bias created because older fossils do not persist until the present as readily. The fairly good agreement of the two curves shows this bias is not very important. From Foote (2000).

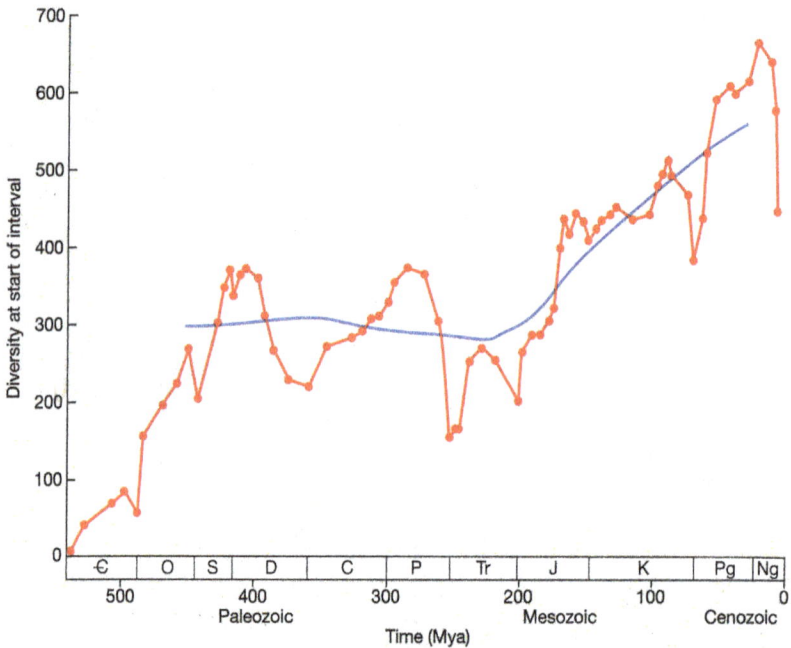

Figure 2.2. Numbers of skeletonized marine animal genera through geologic time, corrected for biases such as temporal differences in rock volume and the fact that more recent fossils are better preserved. The smooth curve is a running average from the late Ordovician to the mid-Cenozoic. The graph shows a steady increase in diversity from the Permian-Triassic mass extinction until recent times even when one corrects for biases. In fact, the increasing trend is even stronger after this extinction than before it. From Foote (2010).

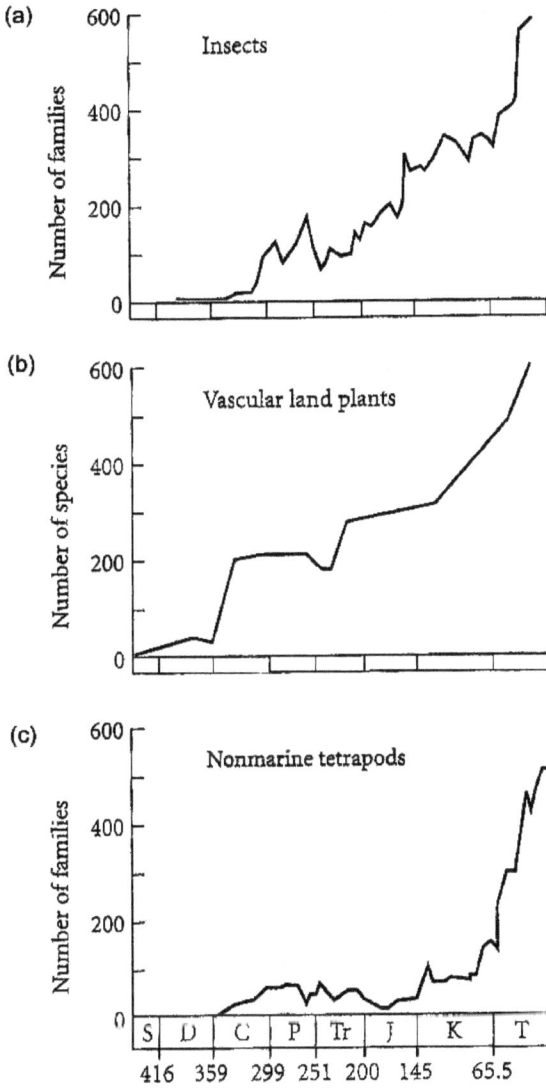

Figure 2.3. Since their origination, changes in the number of known: (a) insect families. From Labandeira and Sepkoski (1993). (b) Vascular land plant species. From Benton (1990). (c) Nonmarine tetrapod vertebrate families. From Benton (1990).

Figure 2.3 shows the number of insect families, vascular land plant species, and nonmarine tetrapod vertebrate families from their origins until recent times. Insects and plants show the basic increase and plateaus, and exponential increase without a plateau since the Permian-Triassic mass extinction, giving further evidence that this pattern holds for a wide variety of taxa. Nonmarine tetrapods do not show an exponential increase until after the Cretaceous-Tertiary extinction, the last of the mass extinctions, which killed off the dinosaurs about 66 mya but, interestingly, the increase is exceptionally steep after this extinction. And it appears that nonmarine tetrapods have not even come close to filling all the niche space available to them on the planet. This does not mean niches remain unfilled for long time periods, because the increase and filling of niches since the Cretaceous-Tertiary extinction is rapid and exponential. And the general increase in terrestrial diversity since the Permian-Triassic mass extinction would have created many new niches that nonmarine tetrapods could have diversified into and filled. The basic pattern of an exponential increase in diversity without a plateau after the Permian-Triassic mass extinction until the present holds in all three of these groups, though the nonmarine tetrapod increase is delayed compared to the others.

The figures all show that life diversified exponentially after the beginning of higher life at the Cambrian explosion and after each mass extinction. Exponential growth of species (or other groups) implies that changes in diversity are guided by a first-order positive feedback, whereby more species (or other groups) create more descendant species (taxa). There may also be second-order positive feedback whereby species number increases with increasing complexity of community structure. This complexity is created by organisms. An example is trees of a rainforest providing a three-dimensional habitat for animals, plants, and fungi. The species that are created provide niches and habitats for additional species. Both first- and second-order positive feedbacks with respect to increases in numbers of species are examples of the ABH.

But why did global biodiversity increase exponentially without ever leveling off after the third mass extinction? One would expect it to plateau after all mass extinctions. We do not know the answer, and research is needed to answer this important, fascinating question. However, we have some plausible explanations.

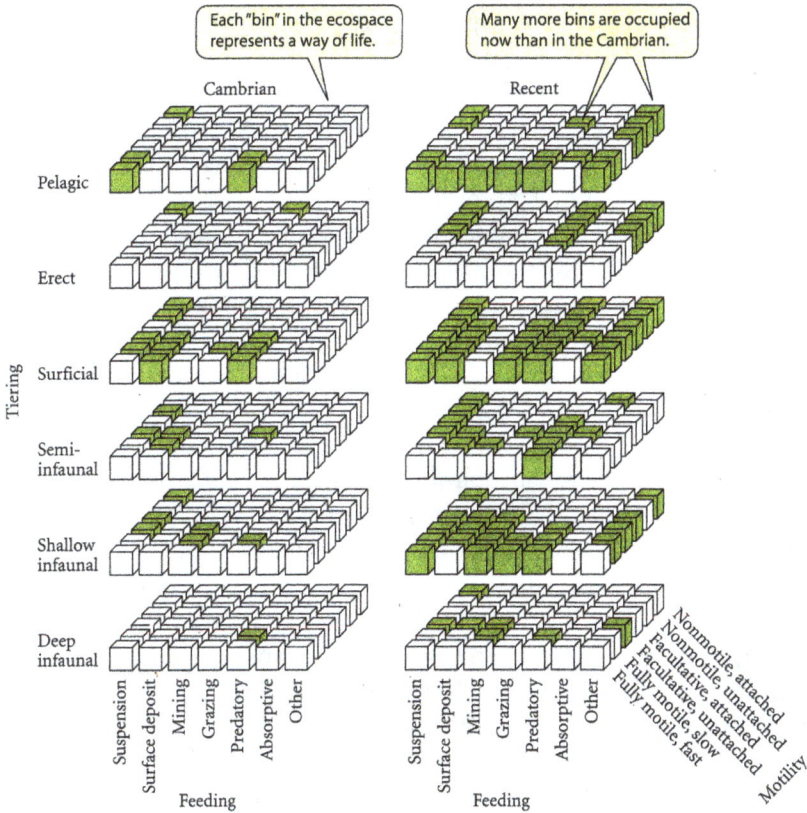

Figure 2.4. Employment of ecological space by marine animals in the Cambrian compared with the Recent (present time). Each layer represents the vertical space used by the animal groups, from pelagic to deep infaunal (deep in the sediment). Cubes from left to right within each layer represent different feeding niches, and those from front to rear represent different motilities. Green indicates the modes of life that are used; white indicates that they are not used. Far more modes of life are employed now than in the Cambrian. From Bush and Bambach (2011).

In the case of skeletonized marine organisms, an increase in use of ecological space provides at least a partial explanation. Figure 2.4 shows that marine animals have a significantly greater variety of feeding niches, habitat types, and motility at the present time than during the Cambrian. This would have been accomplished by active colonization of new habitats, selection, and use of resources and habitat created by other species. The origination of new forms and key adaptations, allowing diversification and employment of more modes of life, would have played a role. This may have kept increasing at a more or less constant rate since the Cambrian, possibly accounting for some of the increase in diversity observed since the Permian-Triassic mass extinction in the sea. Also, a major part of the explanation is that when new forms arise, they provide niches, habitat, or both, for several other species, allowing their appearance. This is true both on land and in the sea.

Another plausible explanation for why diversity has increased continuously since the Permian-Triassic mass extinction is several key innovations appeared during this period. Gymnosperms (conifers, such as pine trees) became dominant and diversified after the extinction, providing three-dimensional habitat for animals. Gymnosperms are also the ancestors of angiosperms (flowering plants), whose appearance was an evolutionary breakthrough that led to a great diversification into many species. The ancestors of angiosperms diverged from gymnosperms in the Triassic, about 245 to 202 mya, and the first angiosperms appeared about 140 mya. They diversified extensively until they were widespread by 120 mya, and have continued to diversify since, and are now 64 orders, 416 families, about 13,000 known genera, and about 300,000 known species. They provided three-dimensional habitat to animals, and created and enriched soil. Angiosperm ancestors coevolved with insects until angiosperms appeared, then insects and angiosperms greatly diversified. Then angiosperms coevolved with other pollinators, and the angiosperms and these pollinators radiated tremendously as a result. These other pollinators included birds, bats, and primates. Then other species were able to evolve to exploit the numerous niches created by the new flowering plants and their pollinators. Insects aided diversification of numerous animal taxa by being pollinators; seed dispersers; decomposers; aerators of the soil; food sources; regulators of populations by being predators, herbivores, and disease vectors; and other mechanisms. The radiation of angiosperms and insects helped birds and mammals diversify tremendously. Birds and mammals then provided many novel niches for many other species, allowing a large additional increase in the number of species and some groups above the species level.

Roots of agiosperms broke down rocks and soil, releasing nutrients that were washed to the sea, where they aided coccolithophores, dinoflagellates, diatoms, and foraminiferans, allowing them to replace less complex forms and become today's four major groups of phytoplankton. This added complexity to the sea. These four phytoplankton taxa greatly diversified. Since phytoplankton are the base of the major food webs of the sea, their diversification promoted diversification of all levels of the marine food webs, from zooplankton to large fish and marine mammals. Angiosperm pollen also went into the sea, providing nutrients to organisms there. A good deal of pollen ends up in the sea, including on the seafloor, as food. Pollen was found in a deep-sea trench 35,000 feet below the Pacific Ocean's surface.

Grasslands released silica to freshwater ecosystems in the Miocene Epoch, which occurred about 23 to about 5 mya. Diatoms are one-celled phytoplankton that need silica for their hard, protective cell walls because these are made almost entirely of silica (Kidder and Gierlowski-Kordesch, 2005). Miocene volcanism also likely released nutrients that aided grasslands and caused them to release more silica to freshwater ecosystems. Volcanism also probably helped freshwater ecosystems and the diatoms inhabiting them by directly providing them with nutrients. It was mainly the grasslands, however, that greatly increased freshwater diatom diversity by releasing silica at this time (Kidder and Gierlowski-Kordesch, 2005).

Cazzolla Gatti (2011) suggested that species themselves generate higher biodiversity by creating niches for other species and thus increasing the niches available in a given ecosystem, in his Biodiversity-related Niches Differentiation Theory (BNDT). Cazzolla Gatti et al. (2017) argued that ecosystems can be viewed as an emergent autocatalytic set in which one group of species enables the existence of other species by creating niches for them. An autocatalytic set is a collection of entities, each of which can be created by other entities within the set, allowing the set to catalyze the increase of the number of its entities (in this case, species). They showed that biodiversity can be considered a system of autocatalytic sets. They argued that such a view explains why so many species can coexist in the same ecosystem.

There is no question that species provide niches for other species, providing sources of food or habitat. All multicellular organisms provide habitat for a diverse microbiome. Trees and other organisms provide habitat for other species. All hosts provide a habitat for their parasites. All prey provide niches for their predators. Provision of niches by species for other species helps explain why life generates higher biodiversity, and provides both evidence for and one of the mechanisms for the ABH. We only understand a small part of the entire explanation for the strange history of Earth's diversity at the time of this writing, but the best explanations are consistent with the ABH.

References

Dasmann, R. F. (June 1965). *The Destruction of California*. Macmillan Pub. Co.: New York, NY.

MacKinnon, J. B. (2013). *The Once and Future World: Nature As It Was, As It Is, As It Could Be*. Houghton, Mifflin, Harcourt: Boston, MA, New York. ISBN 978-0-544-10305-4

Erwin, T. L. (1982). Tropical forests: their richness in Coleoptera and other arthropod species. *The Coleopterists Bulletin* 36: 74–75.

Mora, C., et al. (23 Aug. 2011). How many species are there on Earth and in the ocean? *PLoS Biology* 9 (8): e1001127. doi:10.1371/journal.pbio.1001127

Barrowclough, G. F., et al. (23 Nov. 2016). How many kinds of birds are there and why does it matter? *PLOS ONE*. doi: 1371/journal.pone.0166307

Locey, K. & Lennon, J. (2016). *Scaling laws predict global microbial diversity*. PNAS USA. doi: 10.1073/pnas.1521291113

Leigh, E. G. Jr., & Vermeij, G. J. (29 May 2002). Does natural selection organize ecosystems for the maintenance of high productivity and diversity? *Philosophical Transactions of the Royal Society* B 357 (1421). doi:

10.1098/rstb.2001.0990. Theme issue: The Biosphere as a Complex Adaptive System.

Brayard, A. (15 Feb. 2017). Unexpected Early Triassic marine ecosystem and the rise of the Modern evolutionary fauna. *Science Advances* 3 (2): e1602159. doi: 10.1126/sciadv.1602159

Parrott M. L., et al. (23 Sept. 2019). Eat your heart out: choice and handling of novel toxic prey by predatory water rats. *Australian Mammal.* doi: 10.1071/AM19016

Sepkoski, J. J., Jr. (1984). A kinetic model of Phanerozoic taxonomic diversity. III. Post-Paleozoic families and mass extinctions. *Paleobiology* 10: 246–267.

Foote, M. (2000). Origination and extinction components of diversity: general problems. In Erwin, D.H. & Wing, S. L. (eds.), Deep Time: Paleobiology's Perspective, pp. 74–102. *Paleobiology* 26 (S4), supplement. doi: 10.1017/S0094837300026890. Published online by Cambridge University Press, 26 Feb. 2019.

Benton, M. J. (1990). The causes of the diversification of life. In Taylor, P. D. & Larwood, G. P. (eds.), *Major Evolutionary Radiations*, pp. 409-30. Clarendon Press, Oxford, U. K.

Foote, M. (2010). The geological history of biodiversity. In Bell, M. A., et al. (eds.), *Evolution Since Darwin: The First 150 Years*, pp. 479–510. Sinauer, Sunderland, MA.

Labandeira, C. C. & Sepkoski, J. J., Jr. (1993). Insect diversity in the fossil record. *Science* 261 (5119): 310–315. Bibcode: 1993Sci...261..310L. doi: 10.1126/science.11536548. PMID 11536548

Bush, A. M. & Bambach, R. K. (2011). Paleoecologic megatrends in marine metazoa. *Annual Review of Earth and Planetary Sciences* 39: 241–269.

Kidder, D. L. & Gierlowski-Kordesch, E. H. (April 2005). Impact of grassland radiation on the nonmarine silica cycle and Miocene diatomite. *PALAIOS* 20 (2): 198–206. doi: 10.2110/palo.2003.p03–108

Cazzolla Gatti, R. (2011). Evolution is a cooperative process: the biodiversity-related niches differentiation theory (BNDT) can explain why. *Theoretical Biology Forum* 104 (1): 35–43.

Cazzolla Gatti, R., et al. (2017). Biodiversity is autocatalytic. *Ecological Modelling* 346: 7.

CHAPTER 3. LIFE REGULATES THE ATMOSPHERE'S GREENHOUSE GAS LEVELS AND THE EARTH'S TEMPERATURE

Terminology

This chapter will start by defining some terms necessary for the layperson to understand this and subsequent chapters. The scientist or layperson familiar with the carbon cycle and food webs can skip these paragraphs. *Primary producers* are organisms that acquire their energy via photosynthesis from sunlight and nonliving sources, such as nutrients in the soil, or, if they are aquatic, the nutrients in the water. Plants are primary producers. *Phytoplankton* are small, unicellular organisms that live in the shallower portions of the ocean. They are primary producers and are at the base of most oceanic food webs. A *food web* is a system of nutrient flow that starts with primary producers that are eaten by herbivores that are in turn eaten by predators, with the last two eaten by scavengers after dying. All of these organisms are broken down by *decomposers* when they die. Decomposers are part of the food web. They recycle the nutrients in the organisms they decompose. These nutrients are then again used by primary producers, making the food web a cycle. A food web is not called a food chain because there are many species of primary producers, herbivores, predators, and decomposers, so it is a branched web, not a linear chain. Nature has *food webs*, not *food chains*. An example of a food web is as follows. Phytoplankton of many species produce food (carbohydrate) by photosynthesis, and the phytoplankton are eaten by several species of very small organisms called zooplankton, which are eaten by many species of small invertebrates such as shrimp and krill, which are eaten by many species of small fish, which are eaten by a few species of large fish (such as sharks and swordfish), and then all of the animals are eaten by scavengers after dying, and all of the species mentioned here are decomposed and recycled by decomposers such as bacteria and fungi after their death, in the sea. (Krill are small, shrimp-like crustaceans found in all the world's oceans that feed on plankton, are *extremely* abundant, and are a major food source for whales, seals, penguins, squid, and some fish.) Some phytoplankton are prokaryotes, some are eukaryotes. *Cyanobacteria* are bacteria, and hence are prokaryotes, so they do not have chloroplasts.

But they do carry out photosynthesis. Prokaryotes that carry out photosynthesis do so without chloroplasts. *Oxygenic photosynthesis* is photosynthesis that produces oxygen. Oxygenic photosynthesis is the chemical reaction that uses the energy of sunlight to convert carbon dioxide and water into carbohydrate (food) and oxygen. Land plants, seaweeds, phytoplankton, and cyanobacteria are examples of organisms that carry out oxygenic photosynthesis. It is important because it produces carbohydrates, which are a major food source for organisms, and oxygen. Some bacteria carry out *anoxygenic photosynthesis*, which uses the energy of sunlight to produce food, but uses hydrogen sulfide instead of water, and produces sulfur instead of oxygen. Hence, anoxygenic photosynthesis is photosynthesis that does not produce oxygen. *Respiration* is the reverse reaction of oxygenic photosynthesis, and is carried out by the vast majority of organisms. It uses carbohydrates and oxygen to produce carbon dioxide, water, and energy. It is important because it is a source of energy that allows organisms to survive and thrive, and makes higher life possible. Oxygen is important because it is used to produce energy used by many organisms, including higher organisms, via respiration. Without respiration and sufficiently high atmospheric levels of oxygen to allow organisms to carry it out, higher life would not be possible.

The Carbon Cycle and Negative Feedback Loops

After the sun heats the Earth, some of the heat radiates back into space, so the Earth loses some of the heat it acquired from the sun. Greenhouse gases trap the heat leaving our planet and reflect it back to Earth, and so raise its temperature. Carbon dioxide and methane are the main greenhouse gases, and methane is about 80 times as powerful as carbon dioxide at warming the planet in the first 20 years after it is emitted. Yet carbon dioxide is the major greenhouse gas because it is an effective one, and there is much more of it in the atmosphere than methane. Both carbon dioxide and methane contain carbon. Therefore, if carbon is buried or otherwise removed from the atmosphere, the amount of carbon dioxide, methane, or both will be reduced in both the atmosphere and in rivers, lakes, and the sea. If carbon is added to the system, there will be more carbon dioxide and/or methane in the air and in water. Carbon naturally moves through the air, water, organisms, and the ground. It gets buried and released back into the atmosphere. This constant movement of carbon through all of these places is called the carbon cycle. Carbon sources add carbon to the biosphere and atmosphere. The biosphere is the part of the Earth occupied by life. They add carbon by such mechanisms as volcanoes, movement of the continents over the seafloor's calcium carbonate and organic sediments (both of which contain carbon), and spreading of the seafloor. A carbon sink is anything that removes more carbon from the atmosphere than it releases to it. This carbon is stored in living organisms and the Earth, including limestone and the seafloor. This process of removal of carbon from the biosphere and storing it elsewhere is called carbon sequestration, and includes weathering and burial. It lowers the temperature because it decreases the amount of greenhouse gasses in the biosphere. Weathering is the breaking down of rocks, soil, minerals, wood, or even artificial materials through contact with the Earth's atmosphere, water, and biological organisms. In it, carbon dioxide is combined with minerals in chemical reactions. The resulting material is then carried via rivers and creeks to the sea, where it falls to the ocean bottom and is buried. Weathering is a key step in sequestering carbon, removing it from the biosphere. Weathering is the main nonbiological way carbon is removed from the atmosphere. Weathering is done biologically too, which also sequesters carbon, and this will be discussed later. Nonbiological effects on atmospheric carbon can be profound. For example, when one continental plate slides over another, this buries carbon, sometimes in great quantities. Volcanoes return this carbon to the atmosphere in the carbon cycle. Nonbiological processes can also send large amounts of greenhouse gases into the atmosphere. Carbon dioxide could potentially build up to high levels in the atmosphere if chemical and physical processes were the only carbon sources and sinks. It is added in great quantities annually, to the tune of 280 to 360 thousand kilograms (about 309 to 397 tons), through gases leaving the Earth's interior, such as from volcanic eruptions and mid-ocean ridges. Also, water vapor is the gas that volcanoes emit the most, and it is a greenhouse gas, although it only stays in the atmosphere for short time periods, tending to rapidly become snow or rain. Volcanoes also sometimes emit methane. Some volcanoes emit mainly particles that reflect solar radiation into space and thus cool the planet, but most emit more greenhouse gases than particles, and so warm it. There have been times in geological history when massive volcanism spewed tremendous quantities of greenhouse gases into the air and greatly heated the Earth. Over time scales of tens to hundreds of millions of years, greenhouse gas levels in the air may vary due to natural perturbations. Yet there have been billions of years of life-friendly temperatures that have stayed within a fairly confined range, which required a tight balance between carbon sources and sinks over these timescales. Zeebe and Caldeira (2008) presented direct evidence for such a close mass balance of surface carbon on shorter timescales than these. They concluded that, over the last 610,000 years, the net imbalance between carbon inputs and outputs from the surface environment was less than 1–2%. This means that carbon on Earth's surface has been kept at a very constant level for over the last 610,000 years. This is likely largely because under normal circumstances, both nonbiological and biological factors keep the carbon levels stable by negative feedback. Feedback is a process whereby a cause is modified by its effect. Negative feedback is a process whereby one factor, call it factor A, causes another factor B to increase, and factor B causes factor A to decrease. There can be more factors in the loop. So if A increases B, which increases C, which comes back and decreases A, there is negative feedback. Because this process cycles, it is referred to as a *negative feedback loop*. Negative feedback loops keep the system stable, keep the level of greenhouse gasses in the atmosphere and the temperature stable, and are beneficial to life at intermediate levels. Normally, nonbiological factors stabilize carbon dioxide levels and temperature by negative feedback, by the following mechanism. When there is more carbon dioxide in the atmosphere, the temperature increases by the greenhouse effect. Air can hold more water vapor at higher temperatures, more water will evaporate and eventually fall somewhere as rain, which increases weathering, which will remove carbon dioxide from the atmosphere. When there is less carbon dioxide in the atmosphere, temperature decreases, so the atmosphere can hold less water, so rainfall decreases, weathering that removes carbon dioxide from the air decreases, carbon sequestration decreases, and so the carbon dioxide level and temperature increase. This mechanism stabilizes carbon dioxide levels and temperature, keeping them at intermediate levels favorable to life.

Positive feedback or a *positive feedback loop* is destabilizing, and occurs when A causes B to increase and B causes A to increase. Again, there can be more than two agents. If A increases B, which increases C, which loops back and increases A, there is a positive feedback loop. If carbon dioxide levels in the air become too high, positive feedback can replace the negative feedback. For example, humans have caused increased carbon dioxide in the air, warming the temperature, and causing the ice over the sea on the coast of Greenland to melt. This exposes the sea, which is darker than the ice that has melted and disappeared. Darker surfaces absorb heat, while lighter surfaces reflect heat. So melting the ice and exposing the darker sea causes more heat to be absorbed, temperatures get hotter, and more ice melts. This causes the sea to be even darker, causing it to absorb yet more heat, leading to higher temperatures, melting yet more sea ice. The positive feedback loop could keep going until all the sea ice is gone. This does not contradict the fact that there is negative feedback on carbon dioxide levels, as discussed earlier, because the positive feedback only takes effect when greenhouse gas levels and temperatures reach extreme levels. Positive feedback can lead to catastrophe. Although nonbiological factors in normal circumstances stabilize greenhouse gas levels and temperature via negative feedback, they can cause positive feedback and destabilize the system in rare circumstances. This is because large, uncommon perturbations caused by catastrophic nonbiological events such as massive volcanism or large meteor strikes, or by unnatural phenomena such as humans adding too much greenhouse gas to the air, can throw the temperature to extremes.

Both the Gaia Hypothesis and the ABH (Pachamama Hypothesis) would be best supported if there were stabilizing negative feedback of greenhouse gas levels and temperature by life. In fact, life has a homeostatic (stabilizing) mechanism involving negative feedback that regulates both carbon and temperature (Lovelock and Whitfield, 1982). Here is how life employs negative feedback to keep greenhouse gas levels and temperature stable, and at intermediate levels favorable to life. Bear in mind in this discussion that burial of carbon removes the two main greenhouse gases, carbon dioxide and methane, from the air, since carbon makes up a large part of both of these gases. If the concentration of carbon dioxide in the atmosphere rises, plants and phytoplankton carry out photosynthesis faster.

Life Regulates the Atmosphere's Greenhouse
Gas Levels and the Earth's Temperature

Photosynthesis consumes carbon dioxide, removing it from the atmosphere. Laboratory experiments have demonstrated that if there are higher levels of carbon dioxide in the air, plants and phytoplankton take up more of it for photosynthesis if their growth is not limited by light, water, or nutrients [e.g., Bazzaz, 1990]. They increase in number, and so more die. About the same percent of plants are buried and about the same percent of phytoplankton sink to the seafloor. But because there are more of both when the air has more carbon dioxide, more plants are buried and more phytoplankton sink to the bottom of the sea. Thus, the dead plants and phytoplankton sequester more carbon when they are buried and sink. Therefore, the atmosphere has less of the main two greenhouse gases, carbon dioxide and methane. This lowers the temperature. On the other hand, when carbon dioxide levels are low, plants and phytoplankton absorb less of it, and its atmospheric level increases. And when there is less carbon dioxide and it is cooler, plants and phytoplankton grow more slowly and become less abundant, so less of their biomass and carbon is buried, and carbon dioxide levels and temperatures rise. Biomass is the weight or mass of a living entity, which could be an organism, a forest, or all the phytoplankton in the sea, for example. Thus, life acts as a thermostat that regulates carbon dioxide and temperature to within a range favorable to it, by negative feedback. And, as will be discussed later, animals called salps are, like phytoplankton, important in negative feedback and regulation of carbon in the sea. In a book, Lenton and Watson (2011) discussed various mechanisms by which negative feedback has stabilized the Earth's systems, including its temperature, throughout its history, many of these mechanisms driven by life.

Both these feedback systems have limits, and are far from perfect. In negative feedback driven by biology, if greenhouse gas levels or temperature increase excessively, plants and phytoplankton grow more slowly or even do not do well, and the negative feedback breaks down. And nonbiological factors can destabilize the system for long periods. For example, volcanism could greatly increase, adding tremendous amounts of carbon dioxide to the air and driving the temperature up enormously. It is thought that the Permian-Triassic mass extinction of approximately 251.9 mya, the largest extinction in Earth's history, resulted from volcanism belching so much carbon dioxide into the air that temperatures soared so high that the frozen methane in the sea in the form of what are called methane hydrates was melted, releasing great quantities of methane into the air, making the temperature exceedingly hot. This melted more methane hydrates, adding yet more methane to the atmosphere, making the Earth even hotter, in a positive feedback loop, until the planet was too hot for most life to exist.

Additionally, the stabilizing biological negative feedback mechanism can be disrupted by large external inputs that throw the system to an extreme. In such cases, life takes part in a destabilizing positive feedback loop. For example, human-induced global warming has led to melting of the permafrost, resulting in the release of greenhouse gases by organic matter and microbes, in a positive feedback loop involving biology. The positive feedbacks are not permanent, and throughout life's history, negative feedback and stable, favorable conditions for life have always eventually been restored, generally with life playing a major part in this.

With this background, this chapter will make the case for the importance of life in regulating carbon (and hence carbon dioxide and methane) in the atmosphere and thus stabilizing Earth's temperature, major tenets of both the Gaia Hypothesis and the ABH/Pachamama Hypothesis. Although Lovelock (1979) argued that life has regulated and does regulate Earth's greenhouse gases and temperature, the arguments presented for the ABH here are more specific, broad, and thorough, partly because we know more about this topic now than when he wrote about it.

Life is the Main Regulator of Atmospheric Carbon

Figure 3.1 shows atmospheric carbon dioxide levels for about the last 570 million years, or since about 29 million years before Cambrian explosion, when most of the major animal groups first appeared. The fluctuations in carbon dioxide levels show that the negative feedback loops that stabilize them involving both life and nonbiological factors are imperfect. The figure shows that carbon dioxide peaked about 520 mya, and, although there have been both large fluctuations and periods of increase, carbon dioxide levels have decreased to preindustrial levels since then, in a somewhat steady downward trend. It is also true that the average global temperature has fluctuated a good deal, but is lower today than it was 600 mya, and has been decreasing steadily from about 60 mya until preindustrial times. As will be discussed, these two downward trends are mainly due to life's sequestration of carbon, and are highly beneficial to life.

The amount of heat the sun gives off has increased about 30% since life began about 3.5 billion years ago (bya) (ibid.). The heat the sun has provided the Earth has increased gradually and steadily, except for minor variations, until the present, and will continue to do so into the future, with its brightness increasing by 1% every 100 million years. There is also a continuous input of carbon into the air from mid-ocean ridges, volcanoes, and other sources, although this is small compared to the transfer of carbon from the air to the living organisms via photosynthesis (Frings, 2019). For life to thrive, a mechanism to keep Earth warm had to be present when life had just begun and the sun produced much less heat. As time went on, there had to be a mechanism to keep cooling the Earth to counter the continual increase of heat from the sun and addition of greenhouse gasses to the atmosphere from volcanoes and other sources. Both of these were accomplished mainly by life.

With the sun producing such a small amount of heat at life's beginning, when the only life were prokaryotes, Earth would have been an icy, cold wasteland with its seas frozen, unless a heat-generating mechanism existed. Ueno et al. (2006) found evidence that the source of warming needed to keep Earth favorable to life was produced by methanogens. Methanogens are archaea that produce the powerful greenhouse gas, methane. These researchers found deposits that bear methane produced by methanogens in structures in Australia that are about 3.5 bya, showing that methanogens had arisen by this time, warming the Earth by producing large quantities of methane over a 100-year period. Sediments laid down by water show the Earth and its seas were not frozen, and liquid water existed 3.8 to 2.5 bya. In fact, geological and biological evidence suggests that Earth was warm during most of its early history, despite the fainter young sun (Haqq-Misra et al., 2009). Something provided a source of warming to allow early evolution to occur, and it was life in the form of methanogens. Haqq-Misra et al. (2009) pointed out that carbon dioxide levels 40 times higher than preindustrial levels because of volcanism and reduced weathering were present then as well. These high carbon dioxide levels were also needed to maintain sufficient temperatures for life to thrive. In fact, the high levels of methane would have reflected solar radiation into space with a thick haze, keeping the atmospheric temperatures close to freezing levels inhospitable to life, had the carbon dioxide not diluted it. Methane is a powerful greenhouse gas that generally warms the Earth, but at high levels, it can reflect solar radiation into space and lower the temperature considerably. Thus, carbon dioxide produced nonbiologically also played a role in heating Earth. However, without the methanogen-generated methane warming the air, the atmosphere would have been too cold for life to thrive and diversify, even with the high carbon dioxide levels, because of miniscule amount of heat provided by the sun. Atmospheric methane levels would have had to be 1,000 parts per million by volume, combined with levels of carbon dioxide 100 times preindustrial levels, in the middle Archean Eon (3.8 to 2.5 bya), to reach temperatures comparable to those of today. The right combination of carbon dioxide and methane concentrations were needed to keep the planet warm enough for the prokaryotic life at the time to thrive. There are other hypotheses that attempt to account for the heat required for life to thrive at this time, but in my view, they lack evidence as convincing as that of the work by Ueno et al. and Haqq-Misra et al. Thus, life was the primary facilitator of warming for the early Earth, with help from a nonbiological factor. This warming of the planet allowed prokaryotes on the young Earth to diversify into many different species. This is an intriguing example in support of the Pachamama Hypothesis.

Pavlov et al. (2000) found evidence that methane in the atmosphere 2.8 bya, presumably produced by methanogens, could have warmed Earth at this later time, and that its great decline with the rise of oxygen, which combined with it, about 2.5 bya could have triggered the Earth's first widespread glaciation. So methanogens and the facilitation of warming they provided were fundamental to the thriving and diversity of prokaryotes starting at least 3.5 bya and possibly as late as about 2.5 bya, although we do not know if this occurred continuously for this entire time.

The Earth had to be warm enough for complex animals to survive in the

Life Regulates the Atmosphere's Greenhouse
Gas Levels and the Earth's Temperature

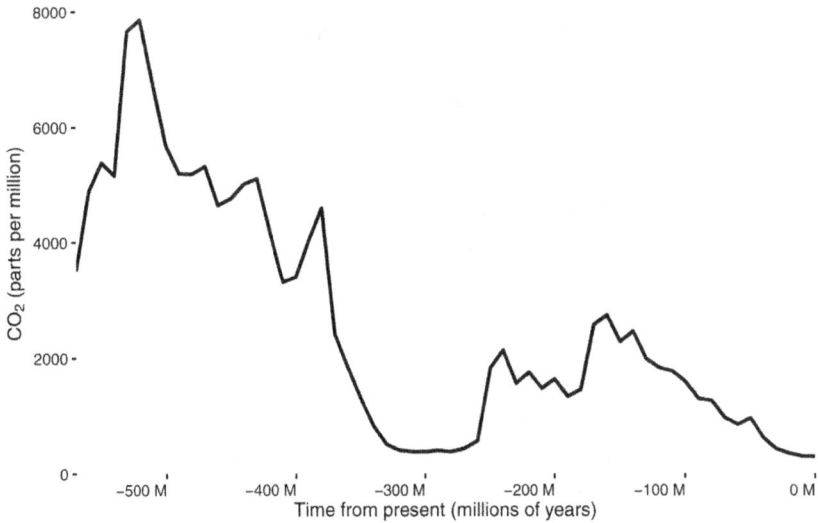

Figure 3.1 Carbon dioxide concentration over the past 570 million years.
Data from Berner and Kothavala (2001).

---

sea at the time of the Cambrian explosion, when almost all the ancestors
of modern forms of animals appeared in the ocean, about 541 to 530 mya.
The sun has increased its heat output by about 6% since this time. And
mid-ocean ridges, volcanoes, and other sources have added greenhouse
gases since this time. The surface of the Earth would have become torrid, far too hot for higher life, including mammals, birds, reptiles, plants,
and fungi, unless something had been removing greenhouse gases from
the atmosphere since about 550 mya, to counter these two sources of heat.
In fact, organisms sequestered carbon for the last 550 mya, regulating the
temperature of the planet and the acidity of the seas and freshwater ecosystems, allowing life to thrive. (High carbon dioxide levels in water increase
its acidity, and this is deleterious to life; this will be discussed in more detail
in section 2.9 of this chapter).

It is not known what percentage of carbon is sequestered because of biological as opposed to nonbiological causes, and the relative importance of

35

the two to the carbon cycle and climate has varied at different times in geologic history. But biological processes have been sequestering carbon for at least 550 million years, still are of crucial importance, and are the major mechanism of reducing greenhouse gases in the atmosphere, seas, and freshwater ecosystems, sequestering significantly more carbon than nonbiological forces have done and currently do. Life has buried immense quantities of carbon, thereby removing it from the atmosphere, maintaining the Earth at temperatures favorable to life.

Life regularly returns carbon to the atmosphere when organisms respire and decompose. Life warmed the planet by adding methane when life had not existed for very long and methanogens were dominant, as discussed above. But since those times, the average amount of carbon sequestered by life has far exceeded the average amount that it has been added by respiration and decomposition.

We know that life, with some help from nonbiological weathering, took carbon dioxide levels from a peak of about 7,000 parts per million in the Cambrian Period (541 to 485.4 mya), to below 500 parts per million in about the middle Carboniferous Period, about 330 mya (see Figure 3.1). This was correlated with a drop in global mean temperature from about 21.5°C (70.7°F), which, for a global average, is absolutely scorching and inhospitable to life, about 550 mya, to about 17°C (62.6° F), which is still hot for a global average, but much more favorable to life and biodiversity, about 330 mya. For perspective, the World Meteorological Organization estimates the Earth's annual mean temperature was about 14°C (57.2°F) between 1961 and 1990. Tremendous volcanism shot carbon dioxide levels to 3,000 ppm at the end of the Permian, about 251.9 mya. Global mean temperature spiked to a torrid 22°C (71.6°F). Since then, life has brought atmospheric carbon dioxide and methane down consistently and gradually over a period of almost 250 million years, to the preindustrial levels of what are now thought to be about 260–270 parts per million, resulting in a mean temperature of about 17°C (62.6°F), by about 2 mya. By this time, our Hominid ancestors were present. The changes in greenhouse gas levels and temperature were not linear, since nonbiological factors and variations in the amount that life sequestered carbon through time influenced these variables. Still, the profound effect of life in molding the atmosphere and temperature to favor life and biodiversity is clear and of great magnitude. This chapter will now explain how this feat of ecosystem engineering by organisms was done.

The sequestration of carbon by life lasts for varying lengths of time, depending on the organisms, conditions, and specific mechanism. All of the processes I discuss here sequester carbon for sufficiently long times to affect the Earth's temperature for a long enough time to make a difference. In some cases, the carbon is not sequestered for a long time in a given generation, but the process continues for a long enough time to be significant. Time scales can range from thousands to millions of years.

The biological pump is the sea's biologically-driven sequestration of carbon to the ocean's interior and seafloor sediments. It removes carbon from the atmosphere. Carbon is carried from land into the sea via rivers. The sea removes carbon dioxide and methane from the system through the burial of carbon in the seafloor, thus regulating atmospheric temperature and the ocean's acidity. Every year, the biological pump transfers about 11 billion tons of carbon to the ocean's interior. This is equivalent to more than 66 million blue whales, or about the mass of all nonhuman land mammals on Earth, or approximately twice the mass of all of the people in the world. This takes carbon out of contact with the atmosphere for several thousand years or longer. If it were not for the biological pump, atmospheric greenhouse gas levels would be about 400 parts per million higher than they are today. This would make the Earth so hot that diversity would be much lower than it is at present. Biology accounts for most carbon sequestration in the ocean (Morse and Mackenzie, 1990).

A great deal of carbon is buried in the seafloor as a result of a continuous shower of debris falling through the sea. This debris originates in the upper layers of the sea where plankton carry out photosynthesis. It can be seen by scuba divers. It looks somewhat like falling snow, so is called marine snow. It is made up of parts or all of dead or dying animals, seaweeds, plankton, bacteria, archaea, fecal matter, sand, soot, and other inorganic dust. Most of it consists of aggregates of small particles bound together by a sugary mucus exuded as a waste product, primarily by bacteria and phytoplankton. Mucus secreted by zooplankton also contributes to marine snow aggregates (Miller, 2004). Thus, marine snow is mostly, but not exclusively, made up of and generated by organisms. The aggregates sometimes increase in size, sometimes reaching several centimeters in diameter, and travel for weeks before reaching the seafloor. Marine snow often forms as a result of phyto-plankton blooms, which occur when phytoplankton in the sea reproduce profusely and bloom into great numbers. The phytoplankton from these blooms aggregate, which speeds up their sinking rate. This is a big source of loss of phytoplankton from the surface (Lazier, 2006). Most organic components of marine snow are eaten by organisms, from bacteria to filter-feeding animals, within about the first 1,000 meters (about 3,280.8 feet) of their downward journey. There is an entire group of animals at depths between about 100 (about 328.1 feet) and 1,000 meters that eat marine snow. It therefore supports many animal species. Most of what is eaten gets sent back to the air in a relatively short time, because organisms expel carbon dioxide when they breathe and because if organisms decompose without sinking, this emits carbon dioxide into the air. However, a significant amount is never consumed, and circulates in the sea for a long time, and some of this eventually shuttles carbon into the sea's depths and sometimes the seafloor; all this can keep it out of contact with the atmosphere for more than 1,000 years. This is long enough that marine snow certainly qualifies as an effective, natural carbon sequestration mechanism by life. *Marine organisms sequester about 1012 kilograms of carbon per year as marine snow* (Boyd et al., 2019, and references therein). Marine snow has buried huge quantities of carbon in the seafloor throughout the history of life.

Marine viruses are likely major players in the biogeochemical cycles that run the planet (Danovaro et al., 2011). The viral shunt is the process by which marine viruses burst open the cells of one-celled organisms such as bacteria, archaea, and phytoplankton, killing them and the releasing nutrients from their cells and fertilizing the ocean with them. A liter of seawater collected in marine surface waters typically contains at least 100 billion viruses. Viruses of bacteria are called bacteriophage or simply phage. *Marine bacteriophages kill 20 to 40% of ocean bacteria every day* (Fuhrman, 1999; Weinbauer, 2004, Suttle, 2005; Proctor and Fuhrman, 1990).

Life Regulates the Atmosphere's Greenhouse
Gas Levels and the Earth's Temperature

Mojica and Brussard (2015) showed that as much as 80% of the carbon made into carbohydrates by photosynthesis moves through the viral shunt into the pool of dissolved matter in the ocean, while Wilhelm and Suttle (1999) stated that as much as a quarter of the carbon in complex compounds used by life in the sea flows through the viral shunt. This means viral infection is a very important mechanism for recycling carbon and other nutrients in the marine environment. Every second, approximately 1023 viral infections occur in the ocean (Suttle, 2007). This releases vast amounts of nutrients, including nitrogen, phosphorus, iron, and carbon compounds, into the sea. *The nutrients released cause the growth of tremendous amounts of archaea, bacteria, and phytoplankton.* These small organisms that receive nutrients as a result of the viral shunt support the major food webs in the ocean. *Since the viral shunt provides nutrients to the ocean's main food webs, it greatly increases productivity, biomass, and biodiversity in the sea.*

Suttle (2007) pointed out that the shunt very likely increases carbon sequestration, speeding up the biological pump. The carbon-rich compounds are among the most complex compounds released by the shunt, and are not easily ingested, and so tend to sink with greater probability than the other compounds, selectively sequestering carbon (ibid.). In addition to bacteriophage, viruses of eukaryotic plankton appear to increase carbon sequestration (Blanc Mathieu et al., 2019). In fact, some virus-infected phytoplankton sink more rapidly than uninfected ones (Lawrence and Suttle, 2004), making them more likely to be buried at the seafloor (Lawrence et al., 2002), although this has only been examined in a few plankton.

However, the main mechanism of carbon sequestration by the viral shunt is making nutrients available to prokaryotes and phytoplankton. The shunt causes the release of sufficient iron to supply the needs of phytoplankton (Poorvin et al., 2004). Adding iron and other nutrients to the sea increases phytoplankton and prokaryote numbers. The increase in the populations of these organisms means more eventually sink to the seafloor, where they get buried, sequestering carbon. The viral shunt has almost certainly affected prokaryotes and sequestered carbon by this mechanism essentially since the beginning of life, since prokaryotes and viruses were the first life forms. The quantity of carbon sequestered by the viral shunt as well as its effect on the carbon cycle is unknown, but likely large.

Stromatolites are vertical columns that reach heights of up to one meter (about 3.3 feet) or more and are formed mainly by cyanobacteria and layers of sandy and pebbly sediments. Cyanobacteria were probably the first organisms to evolve oxygenic photosynthesis. One of the places they grow is in shallow seas. Cyanobacteria produce adhesive compounds that cement sand and other pebbles to form microbial mats. Another layer of cyanobacteria grows over the old one, cementing more sand and pebbles. Then an additional layer of cyanobacteria and sand and pebbles grows over the structure. By this process, they gradually build up in multiple layers over time. The structures produced by these layers are stromatolites. The lower layers store carbon that is kept out of the biosphere. Living stromatolites exist, but are rare today. But stromatolites are a major constituent of the fossil record of the first forms of life on Earth, so might date to over 4 bya. They peaked about 1.25 bya and subsequently declined, so that by the start of the Cambrian Period about 541 mya, they had fallen to 20% of their peak. But they were abundant before the Cambrian. Some fossils that are called stromatolites are not biological, so cannot be used as evidence that life sequestered carbon. But biological stromatolite fossils are abundant enough to support the thesis that they sequestered a good deal of carbon. These fossils store carbon that cyanobacteria removed from the biosphere.

As cyanobacteria photosynthesize, a layer of mucus often forms over mats of their cells. This mucus can trap debris from the surrounding habitat. This debris can be cemented together by calcium carbonate to grow thin laminations of limestone. Limestone is in fact mostly calcium carbonate and hence rich in carbon. So this process sequesters carbon. This process happened for long time periods and over great areas in the geologic past. So stromatolites and bacterial mats sequestered a large amount of carbon in the past.

Life started as prokaryotes at least 3.5 bya. One-celled eukaryotes first appeared between 2.1 and 1.6 bya, and the first multicellular organisms appeared between 1.6 and 1 bya. So, from 3.5 bya or earlier to between

Life Regulates the Atmosphere's Greenhouse
Gas Levels and the Earth's Temperature

1.6 and 1 bya, over half of life's history, one-celled forms and viruses were the only life and hence the only biological mechanism of carbon sequestration. These life forms sequestered a good deal of carbon. This was largely accomplished in the sea, where it happened via the viral shunt, by prokaryotes and phytoplankton sinking, by layering of bacteria in stromatolites, by zooplankton defecation, and by the death and sinking of zooplankton.

Carbon fixation is the process by which simple carbon molecules, mainly carbon dioxide, are converted to more complex molecules by living organisms. The molecules are then used to store energy or as structure for other biological molecules, including DNA, RNA, and proteins. Today, most planetary carbon cycling actually occurs in the deep sea where it is too dark for photosynthesis. Pachiadaki et al. (2017) showed that certain nitrite-oxidizing bacteria are the main organisms that do carbon fixation between approximately 220 to 1,100 meters (721.8 to 3,608.9 feet) below the ocean surface, *fixing 15 to 45% of inorganic carbon in this zone*, in the western North Atlantic! They are the most abundant and globally-distributed nitrite-oxidizing bacteria in the ocean, and the primary producers and base of the deep-sea food webs, meaning they support countless species that eat them and are eaten by other species, which are eaten by still others, and so on up to the top of the food webs. The researchers identified over 30 species of nitrite-oxidizing bacteria in this dark part of the sea. They sequester large amounts of carbon when they sink and are buried, although much less than that sequestered by surface phytoplankton. More carbon is sequestered by animals that eat them directly or are higher up the food web than by the sinking of these nitrite-oxidizing bacteria. That is because the animals higher on the food webs, which include crabs and fish, are much bigger than these bacteria, and some die and sink to the sea bottom during any given time period, and this results in the burial of a great deal of carbon there. Carbon is also sequestered when these animals defecate and their feces sink and get buried in the seafloor. The amount of carbon these bacteria sequester, directly and indirectly, is currently unknown, but large; the researchers say that the bacteria that carry out nitrite oxidation in this part of the sea may have a greater impact on the carbon cycle than previously assumed.

Ever since the first appearance of photosynthetic eukaryotes about 1.5 bya, eukaryotic phytoplankton in the oceans have been among the most important groups of organisms at sequestering carbon. Marine phytoplankton carry out about half of the photosynthesis that occurs on the planet (Baumert and Petzoldt, 2008; Simon et al., 2009), and they do this at an astonishingly rapid rate. Phytoplankton, which in this discussion includes photosynthetic prokaryotes such as cyanobacteria, as well as the one-celled eukaryotes, sequester carbon by three mechanisms. First, they bloom in great numbers, followed eventually by large numbers of them dying, sinking, and being buried as sediments, mainly in shallow seas. The second mechanism is aggregation and sinking of cells, which is more important for small cells, such as cyanobacteria. Finally, some are eaten by zooplankton and plankton feeders, which starts carbon's ascent up the food web. Over a given time period, some of the organisms at each level in the food web, from zooplankton to whales, defecate or die, and some of the feces and carcasses sink to the seafloor, where the carbon in them is buried. A staggering 5 to 12 billion tons of carbon per year is sequestered directly or indirectly in the sea by phytoplankton (Siegel et al., 2016). This is a larger quantity than the amount of carbon consumed every year by all the ecosystems on land combined (Boyd et al., 2019, and references therein). It is over half of the carbon dioxide produced each year by burning fossil fuels.

A population of phytoplankton can double its numbers on the order of once per day. Large populations of phytoplankton, sustained over long periods of time, can significantly lower atmospheric carbon dioxide levels. However, note that photosynthesis by phytoplankton or plants alone would not remove carbon dioxide from the biosphere, because the photosynthesizers respire, and decompose after death. Both processes add carbon dioxide back to the biosphere. In order for carbon to be removed from the biosphere, these organisms must be buried in the sea bottom, or be eaten by organisms that acquire their carbon, then eventually die and get buried, or organisms defecate, resulting in their feces getting buried. Of course, organisms also keep carbon out of the biosphere by storing it in their cells and molecules until they die and decompose. This can be a long time and involve a great deal of carbon in large organisms, such as whales, and can be a tremendous amount of carbon in organisms with extraordinarily high population numbers, such as bacteria and viruses.

There are four major groups of eukaryotic phytoplankton today, and they all have been sequestering carbon for over 185 million years. Watson and Liss (1998) speculated that the overall effect of phytoplankton today is to cool the planet by about 6°C (about 10.8°F). All four groups can carry out photosynthesis. Diatoms comprise one group of these phytoplankton. They are one of the major groups of photosynthetic eukaryotes. There are probably well over 10,000 species of them (Mann and Droop, 1996). Their cell walls are primarily made of silica, rather than carbon. Even so, they sequester large amounts of carbon for the following reasons. Diatoms contain carbon as well as silica. The silica cell walls act as ballast when the diatoms die, so they have more of a tendency to sink and thus get buried than other phytoplankton. There are many of them: living diatoms make up a significant portion of the Earth's biomass, and make up almost half of the organic material found in the oceans. The shells of dead diatoms can reach a thickness of over 800 meters (2,624.7 feet), about a half mile, deep on the seafloor. In addition, being silicon-rich, they cause great quantities of silicon to be buried in the seafloor, and this also helps life considerably. Too much silicon in the environment can harm organisms. This latter benefit to life, the burial of silicon, will be discussed later in this chapter.

Dinoflagellates are another type of marine phytoplankton. There are about 1,555 described species of free-living marine Dinoflagellates, and there are estimated to be about 2,300 total living dinoflagellate species, if you count the estimated number of species not yet discovered or described. There are marine, freshwater, and parasitic dinoflagellates. About half of dinoflagellate species are photosynthetic. Dinoflagellates can form protective shells called cysts and lie dormant within them when conditions are unfavorable, and these contain carbon. The cysts sometimes get buried and fossilized, sequestering carbon. Their cysts are found as certain microfossils from the Triassic Period (about 252 to 201 mya), with a few probable records of them from the Permian Period (298.9 to 251.9 mya) (Fensome et al., 1996). They form a major part of the marine phytoplankton fossils from the middle Jurassic, about 73 mya, to the present day, showing that they sequestered large amounts of carbon because of the great numbers of them settling and becoming fossils on the seafloor over large time spans.

Coccolithophores are almost exclusively single-celled marine phytoplankton that are found in large numbers throughout the ocean's photic zone, the shallowest part of the ocean, where enough light penetrates to allow photosynthesis. Many have a broad distribution. When they die and get buried, coccolithophores can take a great deal of carbon out of the biosphere, largely due to their carbon-rich coccoliths, special calcium carbonate plates that cover and protect them and that are important very small fossils (Prothero, 2004). Iglesias-Rodriguez et al. (2008) found that in at least some circumstances, the species *Emiliania huxleyi* becomes 40% heavier—much denser than water, and thus acquiring a greater tendency to sink—and also more abundant, in waters with higher carbon dioxide concentrations. The same *proportion* of them die and sink regardless of how many of them there are, so a greater *number* of them die and sink when there are more of them, sequestering more carbon. Heavier ones contain more carbon, so sequester more carbon when they die and sink. Significantly, this would allow more carbon sequestration in the presence of high carbon dioxide, regulating both atmospheric carbon dioxide and the Earth's temperature like a thermostat by negative feedback. *E. huxleyi* has population explosions that can be seen from orbit. The blooms are killed in great numbers by viruses, sequestering large amounts of carbon. Coccolithophores have both short-term and long-term effects on the carbon cycle. Short-term, the production of their coccoliths actually adds carbon dioxide. They take up some carbon dioxide through photosynthesis, but blooms release and hence increase carbon dioxide in the short-term. But long-term, their coccoliths containing calcium carbonate, which has carbon in it, sink and become part of the seafloor sediment (Marsh, 2003). The long-term effect greatly outweighs the short-term one, so the net effect of the two is a big decrease in atmospheric carbon and cooling of the Earth.

The 10,000 living and 40,000 fossil species of foraminifera make up the other major group of phytoplankton that has sequestered significant amounts of carbon for long periods of Earth's history. They have a *test*, or shell, commonly made of calcium carbonate or agglutinated sediment particles. The ones with tests of calcium carbonate sequester significant amounts of carbon when foraminifera die, sink, and get buried in the seafloor. The principal source of seafloor carbonate from approximately 150 mya to the present is from calcareous phytoplankton, mainly coccolithophores and foraminiferans. Foraminiferans arose about 150 mya, and have grown in abundance since then. The remains of calcareous phytoplankton, such as coccolithophores and foraminifera, fall to the seafloor, accumulate, and form carbonate-rich sediment there. Sedimentation of carbonates has increased since the appearance of calcareous phytoplankton, largely due to them. Presently more than half the global carbonate deposits are in the seafloor (Milliman, 1974). Much of these are from phytoplankton.

It turns out that the great majority of plankton are mixotrophs, which are hybrid organisms that are both predators and photosynthesizers. A system with mixotrophs interacting in a food web with bacteria and phytoplankton removed 65 grams of carbon per square meter of seawater, compared to 30 grams of it per square meter if only zooplankton that are exclusively predators interacted with bacteria and phytoplankton (Mitra et al., 2016; Mitra, 2018). Thus, combining photosynthesis with predation in one organism increases carbon burial significantly. Diatoms, foraminiferans, and radiolarians are mixotrophs (Flynn and Mitra, 2009; Stoecker et al., 2017). Radiolarians are a large group of protozoa zooplankton found throughout the global ocean that primarily eat smaller plankton, but many have symbiotic photosynthesizers in them.

When the asteroid that caused the Cretaceous-Paleogene extinction event that killed off the dinosaurs struck the Earth about 66 mya, debris from its impact and ash from the fires it caused blocked the sun, and the Earth was dark. Photosynthesis in the sea was greatly reduced, including by the phytoplankton at the base of the ocean's food webs. Gibbs et al. (2020) used marine plankton microfossils and eco-evolutionary modeling to find that coccolithophores saved the marine ecosystem by adapting, with some becoming mixotrophs. The system flipped to a bacteria-dominated quasi-stable state when sunlight was lacking. Coccolithophores became mixotrophs, and ate the bacteria. Some of the foraminifera that survived also apparently became mixotrophs, evolving spines that would have worked with miniature tentacles to capture prey. Mixotrophy allowed the coccolithophores and forminiferans to survive and balance the food webs when photosynthesis was difficult to carry out. These two groups may have kept the marine ecosystem from going extinct. Most of them became solely photosynthetic once again when sunlight returned and the ecosystem was re-established two million years later.

When higher animals had just appeared at the start of the Cambrian Period about 541 mya, there were mollusks, calcareous algae, simple corals with carbon-containing skeletons, and animals called brachiopods that looked roughly like clams and had carbon-containing shells and consist of about 12,000 fossil species. All of these carbon-rich organisms removed a good deal of carbon from the biosphere in that period, many continuing to do so up until the present. This is because they contained a considerable amount of carbon and sometimes got buried in the seafloor after death.

Every autumn in the North Atlantic, billions to trillions of the copepod *Calanus finmarchicus* (a crustacean with no common name), which is about the size of a grain of rice, migrate down to hibernate in winter at ocean depths of about 1,400 to 1,600 meters (about 4,593.2 to 5,249.3 feet). Copepods are a group of small crustaceans found in great numbers in nearly every freshwater and saltwater habitat. They transport carbon in fatty molecules called lipids deep in the sea, where it is released and sequestered via their respiration and defecation, and the death and sinking of some of them. Unlike other components of the biological pump, this so-called "lipid pump" does not strip the ocean surface of a significant amount of nutrients. Copepods are especially important, then, since they are among the few organisms that sequester great amounts of carbon in the sea without removing very many other nutrients from the surface waters where they are needed by photosynthetic organisms. The amount of carbon sequestered by copepods equals the amount sequestered by the sinking of marine snow (Jónasdóttir, et al., 2015). This has occurred for sufficiently long geological time periods to have sequestered a tremendous amount of carbon.

Salps are barrel-shaped, free-floating marine animals (phylum Chordata, which includes all vertebrates; subphylum Tunicata, the subphylum that contains sea squirts) that constantly swim and feed on phytoplankton by filtering them from seawater, in tropical, temperate, and polar seas. They range in length from about one to just over 30 centimeters (about 0.39 to just over 15.35 inches), but chains of them can be several meters long. They are abundant enough to affect the biological pump, and they sequester tremendous quantities of carbon. They transport many thousands of kilograms of carbon per day from the ocean surface to the deep sea. They can bud off clones and can probably grow their populations faster than any other multicellular animal, producing swarms that number in the billions and last for months; they can be even more abundant than krill. Salps have no known predators. If their phytoplankton food is too dense, they can clog and die. And when there is no longer a sufficient quantity of phytoplankton to sustain them, they die in great numbers. Their carcasses often sink and get buried, sequestering carbon. They produce a great deal of heavy fecal pellets that quickly sink, efficiently transporting more carbon to the seafloor than their carcasses. One species, *Salpa aspera* (no common name), in the Mid-Atlantic Bight region, consistently multiplies into dense swarms that last for months. One swarm covered 100,000 square kilometers (about 38,610 square miles) of sea surface, containing trillions of salps the size of a human thumb. The swarm consumed up to 74% of the phytoplankton from the sea surface per day, and the salps' sinking fecal pellets transported up to 4,000 tons of carbon a day to deep water. *Salpa aspera* swim down to and stay at depths of 600 to 2,800 meters (1,968.5 to 9,186.4 feet) during the day, probably to avoid damaging sunlight (Madin et al., 2006). Because of this behavior, these salps may release fecal pellets in deep water, where it is unlikely that feces-eaters would consume them. So the pellets reach the ocean floor with high probability. They feed and reproduce at the sea surface at night, so they transport enormous quantities of carbon from the sea surface to its bottom—between five and 91 milligrams/square centimeter/night (ibid.). And when they die, their bodies also sink rapidly, at 475 meters (1.558.4 feet) per day. Salp fecal pellets sink as much as 1,000 meters (about 3,280.8 feet) a day, and far faster and with higher carbon content, making their efficiency of downward carbon movement much higher, than that of the two other major marine invertebrate grazers, which are krill and copepods. Perissinotto and Pakhomov (1998) found that in the Antarctic Ocean's Lazarev Sea during the southern hemisphere's summer (December to January) of 1994 to 1995, the salp *Salpa thompsoni* (no common name) had dense swarms throughout the marginal ice zone, consuming tremendous quantities of phytoplankton. The resulting carbon flux could have attained levels of up to 88 milligrams of carbon/square meter/day, accounting for the bulk of the vertical transport of carbon in the Lazarev Sea, showing the spectacular amounts of carbon that this species can transport from the photic zone to the seafloor. Significantly, there appears to be some stabilizing negative feedback in the carbon cycle (and hence temperature) with Salpa thompsoni. Increases in its biomass and geographic range have occurred in several areas of the Southern Ocean, often in parallel with a rise in sea-surface temperature during periods of warming anomalies. It thus likely sequesters more carbon when temperatures are high, stabilizing carbon levels and the temperature, at least locally.

Giant larvaceans, which are free-swimming relatives of salps, filter water for feeding at 20 milliliters (about 0.04 pints) per second, a higher rate than salps or any other group (Katija, 2017). They shed their filtering structures of mucus, which have diameters over 1 meter (about 3.3 feet), and these sink to the seafloor. They sequester vast amounts of carbon this way.

Ware et al. (1992) showed that the precipitation of calcium carbonate in coral reefs is accompanied by a shift in the acidity of the water that results in the release of carbon dioxide, making them, surprisingly, sources, not sinks, of atmospheric carbon, in the short-term, releasing 18.14 to 72.57 billion kilograms (0.02 to 0.08 billion tons) of carbon as carbon dioxide annually. This is about 0.4% to 1.4% as much as the annual worldwide carbon dioxide production due to fossil fuel combustion by humans in 1992. However, Kinsey and Hopley (1991) stated that coral reefs act as a sink for 111 billion kilograms (about 122.4 million tons) of carbon per year, the equivalent of 2% output of carbon dioxide by humans when they published, although they agree with Ware et al. that the immediate effect of calcium carbonate precipitation is to encourage the release of carbon dioxide to the atmosphere, causing the reef to be a short-term carbon sink. Additionally, in the long-term, the calcium carbonate skeletons of hard corals have been converted to limestone by the deposition of dead corals during long periods in the geologic past, removing carbon from the atmosphere. This has resulted in extensive limestone deposits in various areas around the globe, such as in the Caribbean. Some fish, such as bumphead parrotfish (*Bolbometopon muricatum*), eat coral and excrete white coralline sand, which is full of carbon that does not easily return to the atmosphere. So coral reefs have played a large long-term role in carbon sequestration.

Since a few thousand years after hard skeletons appeared, reefs have existed. They were most prominent in the Middle Cambrian Period (513-501mya), Devonian Period (416-359 mya), Carboniferous Period (359-299 mya), Late Cretaceous Period (100-66 mya), and all of the Neogene Period (23 mya-present). Not all reefs in the past were formed by corals. In the Early Cambrian Period (541-513 mya), they consisted of calcareous algae and archaeocyathids (small animals with a conical shape, likely related to sponges). These sequestered some carbon. In the Late Cretaceous Period, the dominant reef-builders were a group of strange bivalve mollusks called rudists that went extinct at the Cretaceous-Paleogene mass extinction that killed the dinosaurs about 66 mya. (Bivalve mollusks are a large group that includes clams, mussels, and scallops). The fossilized deposits of their carbon-containing shells in the Arabian Peninsula, North Africa and the Near East, the Balkans, the Caribbean, France, Spain, Greece, Italy, Turkey, Mexico, and the U.S. Gulf Coast represent the sequestration of a large amount of carbon. Phytoplankton, some worms with shells, other bivalve mollusks, and other organisms that make calcium carbonate have added to these deposits, sequestering a great quantity of carbon. All told, the amount of carbon in rock far exceeds the amount of free carbon in the air, water, and soil. Limestone is the largest repository of sequestered carbon dioxide on the planet. It represents a huge amount of carbon life has removed from the atmosphere. A small percentage of this carbon is not due to organisms, since, less commonly, limestone can precipitate directly out of solution without life's aid.

Echinoderms are a large group of animals that include starfish and sea urchins. They incorporate carbon directly from seawater into their skeletons in the form of minerals, the main one being calcium carbonate. They also obtain a great deal of carbon from their food. Lehrato et al. (2010) showed that these animals remove surprisingly substantial amounts of carbon—over a hundred billion kilograms (about 110,231,131 tons) of it per year—from the biosphere, since many get buried in the ocean bottom. *This is more carbon than foraminifera deliver to the seafloor sediments and sequester there.* The amount of carbon that Echinoderms sequester is so immense that Lebrato et al. declared, "Climate models must take this carbon sink into account."

Wilmers et al. (2012) found that sea otters in the Pacific northwest seas greatly aid the growth of kelp because they eat sea urchins, which consume kelp. The sea otters cause much more kelp to grow per unit time, and this removes a great deal of carbon from the system. Over a substantial area, the effect of sea otters on living kelp biomass is an increase in carbon storage of 4.4 to 8.7 trillion grams.

Another significant source of carbon transfer to and burial in the seafloor is the sinking of the corpses of large animals. Hypothesized percentages of the contribution of the sinking of large dead animals to the total carbon that sinks to the deep ocean range from 0.3% (Smith, 2005) to 4% (Higgs et al., 2012). The largest animals that sink to the sea bottom upon death are whales. Whale carcasses sequester tremendous amounts of carbon. The amount of organic material reaching the seafloor from the sinking of a whale carcass creates a pulse equivalent to about 2,000 years of background carbon sinking in normal conditions to the 50 square meters (538 square feet) that a typical whale carcass would occupy (Smith and Baco, 2003).). This not only sequesters great quantities of carbon, but provides nutrients to deep-sea ecosystems. Whales need to breathe at the surface, so they tend to defecate in shallow water. Their nutrient-rich excrement tends to float, so is available to phytoplankton in the shallow sea where there is enough light for them to carry out photosynthesis. This reverses the normal downward flow of the biological pump, sending carbon upward in the sea; this is called the whale pump (Roman and McCarthy, 2010). This increases phytoplankton growth. The feces of krill-eating whales are up to 10 million times as iron-rich as the surrounding seawater (Nicol et al., 2011). This is a major source of the iron needed for maintaining phytoplankton biomass on Earth, and aids their blooms. This is especially important in the Southern Ocean, which is rich in other nutrients, but iron-poor (ibid.). This allows plankton populations to explode. This causes phytoplankton to sequester large quantities of carbon when they die and sink. As discussed earlier in this chapter, phytoplankton are major players in carbon sequestration. The iron from the defecation of merely the approximately 12,000 sperm whales (*Physeter macrocephalus*) in the Southern Ocean helps sequester about 200,000 tons of carbon annually. Whales also indirectly help countless species in the ocean's major food webs, since phytoplankton are the sea's major primary producers, at the bottom of the ocean's major food webs. That is, whale excrement helps phytoplankton bloom, and the phytoplankton are eaten by zooplankton, which are eaten by krill, and so on. Entire food webs of animals of many species are supported in this way. Of course, the animals at all levels of the food web also die and sink, and some of their feces sink when they defecate. Some of these corpses and feces get buried in the seafloor. This sequesters large quantities of carbon. So whale feces and phytoplankton play important roles in sequestering large amounts of carbon, both directly, and indirectly through food webs. Also, some whale excrement sinks, directly causing large quantities of carbon to be buried in the deep sea. Defecation by the Southern Ocean's Sperm Whales alone sequesters approximately 400,000 tons of carbon annually. Before commercial whaling, these whales sequestered approximately as much carbon as an average coal-fired plant produces in an equivalent amount of time. There are now about 8,000 blue whales worldwide, the largest animal ever to exist. There were once more than 200,000 blue whales in the Antarctic Ocean alone. All giant whale species were much more abundant in the past. They sequester large amounts of carbon now because of their size. But they sequestered much more carbon in the past because of their greater numbers before humans hunted them to low numbers.

The shells of bivalve mollusks, such as clams, mussels, and scallops, are made mainly of calcium carbonate, and can be broken down to sand, which stays a long time on the seafloor, or they can be buried in the seafloor. The outer skeletons of crustaceans, such as crabs, shrimp, and lobsters, contain a great amount of carbon. When these animals molt (shed their outer skeletons) or die, the molted skeleton or animal is sometimes buried. They molt several times in their lives. Therefore, bivalve mollusks and crustaceans have sequestered a great deal of carbon in the past, and still do today. Precise figures are not known.

Freshwater ferns can be buried, sequestering carbon in great quantities. There was a period when global mean temperature increased by over 5 to 8°C (9 to 14.4°F) (McInherney and Wing, 2011) and was over 8°C (14.4°F) higher than today. There were turtles, crocodiles, and palm trees in the Arctic. Its precise time and duration are uncertain, but it is estimated to have occurred about 55.5 mya (Bowen et al., 2015), lasting for about 200,000 years. The warming probably was ended by the *Azolla* event. Blooms of a freshwater fern of the genus *Azolla* were abundant near the Arctic Ocean. When they died, many were washed by rivers into the Arctic Ocean, sank, and were buried and incorporated into the sediment, sequestering large quantities of carbon. It is postulated that these ferns were buried over time in such abundance as to throw the world's climate from hothouse conditions to a long-term cooling trend that led eventually to the Pleistocene Epoch glaciation that began 2.58 mya. The *Azolla* event did not cause the Pleistocene glaciation, but rather brought Earth from harsh high temperatures to favorable temperatures for life. Sedimentary layers throughout the Arctic basin contain fossilized *Azolla* (Waddell and Moore, 2006). Evidence indicates the *Azolla* event lasted an estimated 800,000 years (Brinkhuis and Schouten, 2006). And it coincides precisely with a dramatic decline in carbon dioxide levels, which fell from 3,500 parts per million in the early Eocene Epoch (which started about 55.8 mya) to 650 parts per million during the *Azolla* event (Pearson and Palmer, 2000). *Azolla* can draw down 1.5 kilograms of carbon per square meter per year (six tons of carbon per acre per year) (ibid.), which is a huge amount. In the warmth and 20 hours of sunlight during the Arctic summer, it can double its biomass over two to three days. Other factors may have contributed to the cooling, but *Azolla* was a major carbon burial agent and important factor that brought Earth out of the very hot conditions to a more life-favorable climate.

On land ecosystems, rainforests sequester huge quantities of carbon. Much of this is accomplished with the help of phytoplankton, working together in a complex system. Leaves and other organic matter full of nutrients from rainforests enter rivers, which transport the nutrients to plankton in the sea, which greatly increase their numbers as a result of the increased nutrients. Some of the plankton die and sink, burying carbon. Some are eaten and the carbon goes up the food web. When animals higher on the food web supported by the phytoplankton die and sink or defecate and their feces sink, carbon is sequestered in the seafloor. All of this sequesters great amounts of carbon. And as a result of rainforest nutrients increasing the numbers of phytoplankton, the phytoplankton produce a great quantity of oxygen. This is a major source of Earth's oxygen (see Chapter 4). Surface plumes transport freshwater discharged by large rivers hundreds to thousands of kilometers away from the coasts. The nutrients carried by the plumes from the rainforest contribute to enhanced production of phytoplankton in the ocean. Near the mouth of a river, the waters are dominated by typical coastal diatoms that are sustained by the nitrates and silicates in the plume. Far from the river mouth, it was thought that there was insufficient nitrate to sustain phytoplankton. But experiments by Subramaniam et al. (2008) in the Northwestern Tropical Atlantic on water flowing from the Amazon River showed that the dominant species of phytoplankton change as the plume moves and the nitrate runs out, so beneficial nitrogen-fixing cyanobacteria provide the nitrate to sustain phytoplankton far from the coast, at least some of which are different species than the ones near the mouth of the Amazon River. This leads to more carbon sequestration. These researchers calculated the production of phytoplankton supported by nitrate coming from the Amazon into the sea to be about 7.2 million tons per year, and that supported by nitrogen fixation by cyanobacteria to be about 20.4 million tons per year, for a total production of about 27.6 million tons of phytoplankton per year in the plume. The key groups in this great carbon sequestration by *interspecific teamwork* are interacting rainforest trees, phytoplankton, and cyanobacteria. The authors have preliminary evidence that the same processes are at work in other tropical rivers that flow into oceans, such as the Congo in Africa, the Orinoco in South America, and the Mekong in southeast Asia. Rainforest trees also store tremendous quantities of carbon in their leaves, trunks, and roots.

In rainforests, elephants selectively eat small, fast-growing trees with wood of low density, allowing large, slow-growing trees with wood of high density to dominate. The latter store carbon better than the former. Without elephant foraging, billions of tons of carbon would be released into the air annually. At a typical density of 0.5 to one animal per square kilometer, elephant disturbances increase aboveground biomass by about 26 to 60 tons per hectare (a hectare is 10,000 square meters, or 2.471 acres). The extinction of forest elephants would decrease aboveground biomass by 7% in central African rainforests. Biomass is the total mass of organisms, including plants, in a given area. To put this in monetary terms, elephants represented a carbon storage service of $443 billion per year in 2019 (Berzaghi et al., 2019).

Trees sequester great quantities of carbon when buried. Harris et al. (2021) estimate that the world's forests are a net carbon sink of 7.6 ± 49 trillion kilograms of CO_2 per year, even when total carbon emissions from human effects such as deforestation are taken into account. The appearance of rooted plants during the Devonian Period, 419 to 360 mya, almost certainly dramatically lowered atmospheric carbon dioxide levels because their roots weathered rocks. Well-differentiated forest soils had developed by the Devonian Period, and deep-rooted vascular plants, spread to upland areas during this time (Berner, 1997). Both of these developments resulted in the sequestration of great quantities of carbon by life, as tree roots and their symbiotic fungi buried carbon by weathering. Vascular plants are higher plants with true roots, leaves, and stems, although some groups have secondarily lost one or more of these traits. They include the most complex plants, and consist of the clubmosses, horsetails, ferns, pine trees, and flowering plants, including trees. In the Carboniferous Period (359 to 299 mya), lower sea levels allowed forests and swamps to grow on lowlands that would otherwise have been under the ocean. This allowed swamps with large trees to be prevalent, and fallen trees are more prone to be covered with sediment in swamps. So great masses of large trees were buried, with the organic matter containing carbon being converted to coal and oil over millions of years, forming much of today's fossil fuel deposits. The trees of the Carboniferous Period sequestered more carbon from the atmosphere than was removed at any comparable time period in life's history. Trees are among the major mechanisms by which life controlled and still controls atmospheric carbon and global temperature. Lignins evolved in the Ordovician Period (485.4 and 443.8 mya). They are a component of wood. Lignin in wood gave a decisive evolutionary advantage for many trees, protecting them from physical assaults, such as fire, and making them less digestible to animals. It also allowed trees to have bark, which gave them strength, structure, and protection from fire, herbivores, and other challenges. It plays a crucial role in conducting water and aqueous nutrients in plants. It aids disease resistance. Lignin is unique to woody plants. It took many millions of years for bacteria and fungi to evolve the ability to break lignin down. Even today, it is only broken down by a limited number of organisms, such as white rot fungi. Before organisms could break it down, lignin increased in the biosphere, and thus large amounts of organic matter accumulated. Trees sometimes attained heights in excess of two meters (about 6.6 feet) at this time. So plant debris with large amounts of undecomposed lignin was deposited and buried in coal swamps on land, and also in the sea, to which it was transported by rivers, and where it was buried with other ocean-deposited carbon-bearing compounds. Vast coal deposits from the Carboniferous and Permian Periods attest to the great amount of lignin that was buried during these periods. Coal is formed mainly from lignin. Lignin makes up one-quarter to one-half of wood when wood is dry. The evolution of this protective substance and the slow response of organisms to evolve a mechanism to digest it for food caused the removal of vast quantities of carbon from the atmosphere. Hence, biological processes—the evolutionary breakthrough of lignin and a slow evolutionary response to it—contributed substantially to large reductions of carbon in the atmosphere and hence a lowering of temperature from the end of the Ordovician Period to the Carboniferous period and beyond it to the Permian Period. The evolution of lignin also was a key innovation that allowed trees to greatly diversify, creating many new tree species.

From the early Earth of over 4 bya to the first appearance of vascular plants on land about 430 mya, weathering happened mostly nonbiologically, and with negative feedback (Berner, 1992). Since vascular plants appeared, they have sequestered, and still sequester, large quantities of carbon by weathering via their roots and the symbiotic fungi associated with their roots, and by holding soil in place and preventing erosion, thus enhancing the rates of chemical weathering of silicate minerals (for a summary, see Berner, Berner, and Moulton, 2003). The plants give the fungi carbohydrates made from photosynthesis and the fungi help obtain minerals and water for the plants. Plants and their associated symbiotic fungi have increased the rate of weathering of land surfaces (Quirk et al., 2015). The mineralogy of highly weathered soils cannot be understood in the absence of plant activity (Lucas, 2001). Roots and their fungal partners secrete organic acids and other chemicals that break down mineral particles in the soil to free nutrients plants need, principally calcium, potassium, iron, phosphorus, and magnesium. These soil minerals combine with carbon, and some of the resulting compounds go into streams and are transported to the sea, where some are buried, sequestering carbon. This is weathering. And some of these compounds with carbon in them provided by plant and fungal weathering are used by fish, bivalve mollusks, coral, some phytoplankton, and others, to build shells, bones, coral skeletons, and so forth. This takes carbon out of the biosphere and stores it in these structures. This of course reduces the amount of carbon in the air, lowering temperature to a level more favorable to life. Plants and their symbiotic fungi are thus regulate temperature and ocean acidity, and provide nutrients to a great number of species of sea life. Plant root fungi are also important in sequestering carbon in their own right. Careful studies have shown that plants with their fungi dissolve rocks five times faster than would be the case without the fungi, in all major forest types (Berner, 1998). Roots with their symbiotic fungi can penetrate rocks and pry them apart, exposing fresh surfaces, and completely dissolve minerals in them by secreting organic acids (April and Keller, 1990; Griffiths et al., 1996; Berner and Cochran, 1998; van Breeman et al., 2000). The actions mentioned in the last two sentences are forms of weathering and cause a good deal of carbon to be sequestered. Also, dead plants, including trees, and dropped leaves decompose into acids that help chemically break down minerals, causing weathering, which sequesters carbon. Overall, vascular plants accelerate weathering and hence carbon burial by about four to ten times (Ward, 2009). Large terrestrial vascular plants are the most effective at weathering. Vascular plants first appeared during the Silurian Period (about 443 to 419 mya), but large land plants with deep roots, such as trees, did not become dominant until the Devonian Period (about 419 to 340 mya). These plants increased weathering and carbon sequestration immensely compared to the algae, lichens, and mosses that preceded them (Berner, 1998). Certainly lichens, and for the most part mosses, are tiny and slow-growing, and have a small interface with rocks. Therefore, though there is evidence that they weather minerals, they do it very slowly. In contrast, trees have root systems that divide into vast numbers of rootlets and work with extensive symbiotic fungal networks associated with their roots and that act as extensions of these roots, causing a large interface with minerals. They take up nutrients quickly and grow quickly. They break up rocks and soil, aiding weathering. We know trees have built much more soil than lichens, demonstrating that trees accelerate weathering much more than lichens. Three different field studies showed trees accelerate weathering rates over lichens and mosses by factors of eight (Drever and Zobrist, 1992), four (Arthur and Fahey, 1993), and three to four (Moulton et al., 2000). So weathering and hence removal of carbon dioxide from the air was tremendously increased by large vascular plants. Today, about one-tenth of the carbon fixed each year is buried with plants and algae that are washed or blown down from land surfaces into seas, rivers, swamps, and other bodies of water. Most weathering by plant roots and their symbiotic fungi results in carbon becoming bound to minerals, being washed into the sea, and getting buried in the seafloor. But also, much of the carbon from plant weathering builds up high levels of carbon dioxide in the soil (Berner, 1992), where much of it is stored and used by life. Water is recycled by plant weathering as well (ibid.).

The increase in weathering after the spread of land plants drove a 40-million-year decline in atmospheric carbon dioxide, which fell from around 4,000 parts per million by volume (ppmv) to about 1,000 ppmv (Beerling and Berner 2005), during the Carboniferous Period. This rate is as fast as any in the deep geologic past, although it is much slower than the changes during the industrial era, which are an increase of over 2 parts per million/year over the past decade (Stocker et al., 2014). *Without plants and their fungal allies accelerating the chemical weathering of rocks ever since this symbiosis first appeared, the atmosphere would have a carbon dioxide content 15 times as high as today's* (Berner, 2013)! That would make the average temperature of our planet a torrid 10°C (50°F) hotter!

When plants drop their leaves, they form a layer of leaf litter above the soil. This forms a continuous, acidic, moist environment that breaks down minerals, causing yet more weathering and hence carbon sequestration. Organic plant litter from leaves and twigs accumulates in the soil too, where bacteria help the process by decomposing the litter to organic and carbonic acids, providing chemicals that dissolve minerals that can combine with carbon, making compounds that sometimes get buried. At the same time, some of the decomposed plant matter returns carbon dioxide to the air. Trees create rain because their roots take water from the soil, and some of this water goes up through the trunk and to the leaves. Some water exits the leaves and becomes water vapor that later falls as rain. In fact, much of the rain from the Earth's heavily forested areas is created by trees (Shukla and Mintz, 1982). This provides more water to dissolve minerals. The recycling of water as rain by trees causes repeated flushing of the water through the soil, increasing weathering (Berner, 1998). The Amazon rainforest recycles rain that comes from the Atlantic Ocean numerous times before it returns to the sea via the Amazon River and its tributaries. About half of the Amazon's rainfall is from rainforest trees. However, some rainforests, such as those of the Amazon lowlands, are flat and have been exposed to so much rain for so long that their soils are thoroughly leached and their roots no longer weather bedrock. So, they do not help sequester carbon. The soils of these forests have very little nutrients.

Another mechanism by which plants sequester carbon is that some plants form microscopic silica grains in their leaves and stems called phytoliths. This includes all grasses, and some herbs, shrubs, and trees. A small amount of carbon becomes encapsulated in each grain. This encapsulated carbon is very resistant to being returned to the atmosphere, even if the plant burns or dies. Unlike most plant matter, which readily decomposes in soil and returns carbon dioxide to the air, the carbon in phytoliths effectively removes carbon dioxide from the atmosphere for millennia (Parr and Sullivan). The carbon can accumulate in the soil (ibid.). Phytolith carbon is estimated to currently extract 300 million tons of carbon dioxide per year from the atmosphere, storing it in the soil for thousands of years. Carbon sequestration by phytoliths varies with the plant species and subspecies, type of biological community (forest, grassland, etc.), climate, and soil conditions. Since this process has likely been going on since the origin of higher plants, the quantity of carbon sequestered by this mechanism since it started has probably been considerable.

The grasslands of the world are a major carbon sink. A school of thought that challenges conventional wisdom and cannot be discounted, given the evidence, argues that grasslands have persistently contributed to keeping the planet comfortably cool for life since the end of the Eocene Epoch, 34 mya (Retallack, 2001). When grasses die, they decompose, and a large portion of their nutrients are returned to the soil. Carbon from grasses is sequestered and stored underground in this way as organic matter. This organic matter held in the soil is a little-known resource of grasslands. It makes the soil fertile and nutrient-rich, and helps it with water retention. Grasslands thus remove large amounts of carbon dioxide from the atmosphere, and store and conserve it as organic soil nutrient for plants and soil micro- organisms. Grasses also cause weathering, freeing minerals from soil. The minerals combine with carbon, and the carbon combined with minerals is taken to the sea via rivers, and some of this sinks to and is buried in the seafloor, resulting in yet more sequestered carbon. This also supplies nutrients to phytoplankton, making marine ecosystems more productive. And as previously discussed, more phytoplankton means more carbon sequestration.

The mycelium (plural mycelia) is the vegetative part of a fungus that does not make spores for reproduction, and consists of a mass of branching, thread-like, usually white filaments that grow underground. Mycelia of soil fungi make oxalic acid and other acids, which, with the enzymes of the fungi, break down rocks in the first step in soil formation. (Enzymes are proteins that aid chemical reactions). They pull two carbon dioxide molecules from the air to make oxalic acid, adding calcium, to make calcium oxalate, sequestering carbon dioxide. They may make other chemicals called oxalates that combine with carbon dioxide, keeping it in the soil. Fungi also change the soil solution chemistry in a way that causes minerals to dissolve, causing weathering. Fungi bury a great amount of carbon dioxide by these mechanisms.

The above list of organisms that sequester carbon is not comprehensive. Skeletons and shells, which have large quantities of carbon, first appeared in the fossil record about 550 mya and have been getting buried with their carbon ever since. Eighteen evolutionary lines of animals have independently evolved exoskeletons with carbon as a major component. All life has carbon in it. Any organism that gets buried on land or in the sea removes carbon from the ecosystem. All groups of organisms sequester carbon because some individuals in each group are buried when they die, and some of the time their feces are buried (for those species that defecate). All species have carbon in their bodies and their feces.

Life Conserves Carbon for the Biosphere, Even as It Regulates It

Life's sequestering of carbon, though beneficial to life, creates a dilemma. It regulates temperature and keeps acidity in the sea and freshwater systems at levels that are beneficial for life. But it depletes the biosphere of carbon, one of the most important elements to life. Carbon is needed by all life. It is a key structural component of all important biological molecules, including DNA, RNA, proteins, and carbohydrates. It is used by plants and phytoplankton to carry out photosynthesis, which supports most ecosystems on land, in freshwater, and the sea. About 99.6% of all carbon on Earth is stored in rock, and only 0.002% of carbon exists in the biosphere (Berner, 1992, and references therein). When carbon is in rocks, it is not accessible to life. The only nonbiological sources of carbon to life are mid-ocean ridges, volcanoes, and other sources of that nature. There is so little carbon dioxide in the biosphere that if it were removed even a little faster than it is replaced, all carbon dioxide in the air could disappear in less than a million years (Berner and Caldeira, 1997)!

But life has mechanisms to alleviate even this problem, to slow down the loss of carbon, to conserve it and keep it available for future use, even as it keeps it out of the atmosphere, regulating Earth's temperature and the acidity of the ocean and freshwater ecosystems. One of the primary mechanisms of accomplishing this is keeping carbon stored in the system in living organisms themselves. Of course, the organisms are using the carbon as they do this. Large amounts of carbon are stored in viruses, prokaryotes, phytoplankton, giant fungi, trees, coral reefs, the shells of bivalve mollusks (such as clams), the external skeletons of arthropods, in large whales, other large mammals, and other organisms. This keeps the carbon out of the air, regulating temperature, without losing it from the biosphere, keeping it available to life. All living organisms are involved in this solution, some more than others. Terrestrial plants store by far the most carbon of any life form. Trees store tremendous amounts of carbon that is therefore not lost from the ecosystem. They store enormous amounts of it in their trunks, leaves, and roots. *Today, the transfer of carbon from the air to the biosphere via photosynthesis, including trees, other land plants, and phytoplankton, is about 1.2 × 1017 grams of carbon per year*, which is a great deal of carbon removed from the air in a small amount of time (Frings, 2019). The amount that terrestrial plants contribute to carbon storage and temperature regulation *while conserving carbon*, and hence the aid they give to the biosphere and biodiversity, is astounding. The biomass of the biosphere is dominated by land plants. An estimated 99.9% of the total quantity of biomass in the biosphere, is classified as vegetation (Lieth and Whittaker, 1975). Marine photosynthesizers are only 0.2% of the global total (Woodwell et al., 1978). Thus, terrestrial vegetation constitutes almost all of the biomass on Earth. Plant biomass is mainly cell wall materials (Duchesne, 1989). Between 40 and 60% of the total mass of plant cell walls is cellulose (ibid.), which is rich in carbon. Therefore, terrestrial plants store almost all of the biosphere's carbon, most of it in their cell walls, and about half of this is cellulose. This is a vast amount of carbon being kept out of the atmosphere and water. All of this carbon is kept available for use by the biosphere. Forests of the coast redwood (*Sequoia sempervirans*), the tallest tree species on Earth, store among the most carbon of any other forest on Earth. Old-growth coast redwood forests produce more aboveground biomass annually, storing more carbon, than forests dominated by any other species (Sillett et al., 2020).

A surprising 50% of carbohydrate production from photosynthesis by plants is stored underground, where carbon is stored and used in the root system, nourishes the symbiotic root fungi, and feeds soil organisms. Typically, 10% of the other 50% that is aboveground is dropped to the soil as leaves. This leaf litter is decomposed by fungi and bacteria, contributing nutrients to soil life, while returning nutrients to the plants, and keeping some carbon out of the air. Rainforests absorbed great quantities of carbon before humans destroyed large portions of them. Before these human impacts, the Amazonian rainforest stored an estimated 1.5 billion tons of carbon dioxide per year, cooling the Earth considerably, while keeping the carbon available to the biosphere.

Bello et al. (2015) found that the loss of large animals that eat the fruit and disperse the seeds of trees in Atlantic tropical forests led to a great loss of the capacity for carbon storage. This demonstrates that large seed-dispersing animals are very important in keeping carbon out of the atmosphere without burying it, thus ensuring that it is stored, conserved, and available to the ecosystem. This is because they help trees by dispersing the seeds of trees and other plants. They eat the fruit, then fly far away from the tree, and defecate (the seeds are in their feces) or drop the seeds a great distance from the tree. Trees that grow far from their parent tree survive much better than those close to it because they are less subject to herbivores and disease, and do not get shaded out by their parent tree. Large animals that disperse plant seeds are thus important ecosystem engineers whose actions remove greenhouse gases from the atmosphere.

Forests cool the Earth by other means than sequestering carbon dioxide. Tropical forests cool Earth by about 1.5°C (2.7°F.). About two-thirds of this is from the capture and storage of atmospheric carbon. The other one-third is the following three factors, in order of decreasing importance: the uneven canopy of the forest has a cooling effect because it provides an undulating surface that can bump overpassing fronts of air up and away from the forest; the release of water vapor by trees; and the release of aerosols by trees that lower temperatures by reflecting sunlight and seeding clouds. All these mechanisms keep the temperature favorably cool for life without removing carbon from the biosphere.

Boreal forests in Alaska, Canada, Scandinavia, and Russia increase temperature because they temper the reflectivity of the snow, which would decrease temperature. Clearing these forests exposes more snow cover in the winter, decreasing temperature because the white snow reflects heat. But considering the collective effects of forests, the net effect of all of Earth's forests is to lower the average global temperature by about 0.5° C.

Bacteria, archaea, and viruses store great amounts of carbon because of their high numbers. Since there are at least 10 billion microbes and 100 billion viruses in a liter of seawater, tremendous amounts of carbon are kept out of the atmosphere while being conserved for use by the biosphere by these organisms. Although viruses are very small, they are among the most important storers of carbon, containing a huge amount of it because of their tremendous numbers. If all the approximately 1031 viruses on Earth were laid end to end, they would stretch for 100 million light years! Suttle (2005) estimates that *marine viruses collectively contain 200 million grams of carbon*. The average marine virus has only $2 \times 10-16$ grams of carbon, but collectively they lock up the same amount of carbon as do 200 million beluga whales (*Delphinapterus leucas*)! Since viruses are short-lived and turn over quickly, this carbon is relatively quickly available to the biosphere. Thus, viruses are one of the most effective mechanisms for keeping large amounts of carbon out of the atmosphere, while keeping it available to the biosphere. Prokaryotes on the land, in the sea, in freshwater ecosystems, and underground also store spectacularly large amounts of carbon. The carbon content of bacterial cells is roughly 50% of their dry weight.

Marine sediments at the bottom of the ocean are important carbon stores that can hold carbon for millennia. Much of the carbon is delivered to them by marine snow and the sinking of dead organisms and feces. Bottom trawling (fishing by dragging heavy nets across the seafloor) can release tremendous quantities of carbon into the ocean.

Another mechanism of keeping carbon out of the air while conserving it concerns the fact that bacteriophages recycle the carbon in deep-sea sediments, making it available to the marine portion of the biosphere (Danovaro et al., 2005). Phages infect, burst open, and kill huge numbers of prokaryotic cells in the ocean depths, liberating and recycling nutrients, and making them available to other prokaryotes. This is vital to the health and diversity of life in the deep sea (ibid.). The researchers also found that a great portion of deep-sea nutrients liberated by the viruses are carried by animals from the ocean depths to the shallower sea, and contribute substantially to the ecosystems in the photic zone. The photic zone sends nutrients, including carbon, to the deep sea as marine snow and from the sinking of carcasses and feces. Thus, the photic zone near the sea surface and the deep sea exchange nutrients, in a mutually beneficial recycling system. These two spatially separated ecosystems are interrelated and aid each other. Viral attack of prokaryotes in the deep ocean closes the loop in the ocean's carbon cycle, helping to recycle carbon back to the photosynthetic zone in the shallow sea. This is of crucial importance, because it keeps a substantial amount of carbon from being buried under the seafloor, keeping it available to the oceanic life, without adding it to the atmosphere. However, it does increase carbon in the sea, so the acidity of the ocean increases. But under natural circumstances, this increase in acidity is not enough to be a problem.

SAR11 is a bacterium with the smallest known genetic structure of any independent cell. It is the most abundant nonviral organism in the sea, making the total amount of carbon it stores enormous. Globally, there are 2.4×1028 SAR11 cells in the seas, half of which are located in the photic zone. A milliliter of seawater has 500,000 SAR11 cells. Their combined weight exceeds that of all the fish in the sea. On average, the SAR11 clade makes up a third of all the cells present in surface waters and nearly a fifth of the cells in the area between about 200 and 1,000 meters (about 656 and 3,281 feet) deep. In some regions, members of the SAR11 clade make up 50% of the total surface microbial community (Morris et al., 2002)! Thus, they store huge amounts of carbon, keeping it out of the air and sea, but available to the biosphere. Recycling organic matter, when they die and decompose, they provide nutrients needed by phytoplankton. SAR11 is very important in the Earth's carbon cycle. But it produces carbon dioxide and water as end products of its metabolism. And when starved for phosphorus, SAR11 generates methane by breaking down a chemical called methylphosphonic acid to obtain phosphorus (Carini et al., 2014). The amount of methane produced is not known. Pelagiphages are marine bacteriophages that greatly slow the addition of these two greenhouse gases to the air by killing millions of SAR11 cells per second, although some CO2 is added to the sea when the SAR11 cells are decomposed. The bursting of SAR11 cells by the pahage also adds crucial nutrients for phytoplankton growth. The result is more phytoplankton, hence more phytoplankton that die, sink, and get buried in the seafloor. This sequesters a great quantity of carbon. It is likely the SAR11 and phage taken together remove a good deal more carbon than they add, but more study is needed to know this for certain.

The world's soils contain and store more than 2.5 trillion tons of carbon. This is largely from bacteria and fungi breaking down plant matter. Also, plant roots, soil microorganisms, and soil animals store large amounts of carbon in soil for varying time periods, and the amount of carbon and length of time it remains in the soil varies with climate, vegetation, soil texture, and how easily water drains from the soil. Carbon is the main element in soil organic matter. Carbon can be released into the atmosphere in days, or remain in the soil for millennia. Burial and decomposition of plants and animals, animal feces, and falling leaves add carbon to the soil. Most of the carbon stored in the soil is soil organic matter, a complex mixture of compounds containing carbon, consisting of decomposing plant and animal tissue, roundworms, fungi, protozoa (one-celled organisms, which include the *Euglena*, *Paramecium*, and some amoebas), bacteria, archaea, and viruses, as well as carbon associated with soil minerals. All of these organisms and minerals contribute to carbon sequestration and storage in soil. Of course, the ultimate source of most soil carbon is photosynthesis by plants. Many of the soil organisms benefit each other in a symbiotic web. Plants exude carbon they do not require for growth through their roots, feeding soil fungi, prokaryotes, and other soil organisms, which render the carbon stable, keeping it in the soil, and creating humus, the organic component of soil, itself a carbon and energy source for soil microbes and plants. Most of the soil organisms are symbiotic with plants, since they help plants obtain nutrients. Some of the carbon used by plants is not sequestered or stored in the soil, but returned to the atmosphere. Carbon compounds also help give soil its water-retention capacity, structure, and fertility. There is a symbiotic feedback, whereby the species of the soil community benefit the soil, which in turn benefits them. Plants and soil organisms store carbon in the soil that can be used later by the ecosystem. Soil bacteria and archaea, and soil itself, are major carbon reservoirs. Bacteria play a huge but poorly understood role in the carbon cycle. The planet's soils contain more than 2.5 trillion tons of carbon, four times as much as aboveground in all organisms and the atmosphere combined (Hathaway, 2001), much of it stored in soil prokaryotes. Soil microbes are a reservoir that stores carbon, keeping it out of the atmosphere and conserving it. Yet much is still unknown about their net effect on the carbon cycle and climate. There are a number of other ways that life can store carbon in the ground while still keeping it available to life, including organic matter recycling, the formation of humus, plant sugars and substances secreted from roots, previously discussed phytoliths, and carbonates.

Special soil habitats also store and conserve carbon. Organisms store a considerable quantity of carbon by creating peatlands, which cover only 3 to 5% of Earth's land surface, but store a quarter of all soil carbon. And when grasslands bury carbon, it is removed from the atmosphere, but is still available for later use. Grasslands store a great quantity of carbon in the roots of grasses and in soil microbes. And when grasslands cool the planet because of their relatively high reflectivity, they do so without burying carbon and removing it from the biosphere—another mechanism by which life creates temperatures favorable to life while conserving carbon.

There is evidence that coastal habitats, such as mangroves, sea grasses, and salt marshes, store 10 times more carbon in their soils per hectare than temperate forests and 50 times more than tropical forests (Pidgeon, 2009). The long-term carbon-storage capacity of one square kilometer (0.39 square mile) of mangrove forest is equal to 50 square kilometers (19.3 square miles) of tropical forest. Mangroves, sea grasses, and salt marshes are very efficient at burying carbon in sediment, where it can stay for centuries or even millennia. Most of the carbon stays in the soil rather than the plant, so little is released when the plant dies. (Rainforests store much of their carbon in the trees. Of course, much of this carbon is recycled into organisms and not sent into the air when the trees die.)

Seagrass beds are among the most efficient carbon sinks, *absorbing carbon up to 35 times faster than tropical rainforests*, while supporting marine life. Ricart et al. (2015) found that seagrass can absorb carbon to the point that it can reduce ocean acidification at local scales. Over six years, they found that areas with seagrass meadows along more than 600 miles of California coastline were on average less acidic with about 20% more carbon stocks than patches at seagrass-sand edges and bare sediments, even at night, when the plants are not performing photosynthesis. Coastal vegetation accounts for about half of the total carbon sequestration in sea sediments even though it accounts for less than 2% of the ocean surface, largely because more of it is usually below the surface than above it, with some plants going as deep as eight meters (about 26.2 feet). The carbon is stored mainly in the root. Mangroves, sea grasses, and salt marshes also purify water, provide habitat for fish and invertebrates to breed, and protect coral reefs and coastal habitats from storms.

Another mechanism of carbon conservation is by decreasing atmospheric methane. Recall that methane is 80 times as powerful a greenhouse gas as carbon dioxide in the first 20 years after it is emitted. In nature, a negligible quantity of methane is produced by nonbiological processes, but most is made by life. Biological production on land is by methanogens, mainly in wetlands, and by microbes in the digestive tracts of termites and ruminants, which release some of it into the atmosphere. (Ruminants are mammals that include cattle, sheep, antelopes, deer, and giraffes.) In the ocean, methane is produced biologically when microorganisms break down organic matter that settles to the seafloor, including dead fish, invertebrates, and bacteria. However, life strongly limits methane production and accumulation. The oxygen in the atmosphere that was produced by photosynthesis reacted in the past, as it does today, with methane to produce carbon dioxide and water vapor, removing methane from the air. Water vapor is a greenhouse gas, but like carbon dioxide, a much less potent one than methane. Today, each year, oxygen from photosynthesis removes much of the between 531 and 792 million tons of methane produced by methanogens, by reacting with it, without losing any carbon from the system. The result of this reaction is a lowering of atmospheric temperature. Since one carbon atom per molecule of methane is lost and one carbon atom per molecule of carbon dioxide is gained, there is no net loss of carbon.

Also, the microbes that produce methane, called methanogens, cannot grow in the presence of oxygen. It is a poison to them. Thus, the growth of methanogens is limited by oxygen as a byproduct of photosynthesis. Plant roots and their symbiotic fungi break up and aerate soil. Burrowing animals, such as gophers, moles, prairie dogs, and earthworms aerate soil. This gets oxygen to methanogens that live in the soil, limiting their growth.

In marine sediments, methanogen numbers are kept low by microbes that use sulfate, not oxygen, as an energy source. These microbes outcompete methanogens for nutrients. Marine sediments are sulfate-rich. Thus, methane from methanogens accumulates in the ocean for the most part only in the rare areas where sulfate is depleted and oxygen is very low or absent. Dissolved oxygen in the sea from photosynthesis also limits methanogens there.

Bacteria of the genus *Beggiatoa* grow near deep-sea hydrothermal vents, in cold seeps in the sea, in sulfur springs, in mud layers of lakes, and in the root area of swamp plants. Some species oxidize hydrogen sulfide to elemental sulfur for energy, reducing the amount of this gas, which is toxic to most species. Some form colonies, or mats, that support microbes that consume methane. Working together, the species of these communities act as biological filters over methane seeps, limiting the amount of methane that enters the water column and the atmosphere.

In addition, methanotrophs are prokaryotes that consume methane as their source of carbon to obtain energy. Some methanotrophs are bacteria; some are archaea. They play a major role in the reduction of methane released into the air from bogs and swamps, where methane production is relatively high.

In some areas in the deep sea, a great amount of methane is released from fossilized reservoirs. Spectacularly, an estimated 90% of this methane is converted to other compounds by groups of various species of methane-oxidizing archaea and sulfate-reducing bacteria, mainly in marine sediments that have no oxygen. Here, methanotrophs that are archaea that do grow in environments that lack oxygen, unrelated to the methanotrophs that require oxygen, consume 90% of the methane formed, as their food source, preventing its build-up. This produces toxic hydrogen sulfide. But it does not persist. Filamentous sulfur bacteria consume it as a food. So a group of methane-consuming archaea and sulfur bacteria work together, reducing the amount of methane released into the air, without increasing toxic hydrogen sulfide. They can form small aggregates or even voluminous mats. And there is another microbe in the archaea that can use nitrate to consume methane (Haroon et al., 2013).

There are many species of dung beetles; Australia alone has 400 species. They eat and decompose dung. Dung is a source of methane. A study showed that six-day-old cow feces without dung beetles produced five times higher methane emissions per day than those with the beetles. Dung beetles tunnel through dung, allowing oxygen to enter it. This kills the methanogens because they cannot grow in the presence of oxygen (Penttilä et al, 2013). And of course, the oxygen combines with the methane, reducing the amount of it. Reducing the amount of methane and the quantity of methanogens that produce it controls temperature while conserving carbon. The beetles also disperse grass seeds, and fertilize and aerate the soil, and make channels that allow water to soak into it more easily. And soil aeration, just as aeration of dung, reduces the amount of methane by destroying it and preventing the growth of methanogens. The tunneling of dung beetles in the soil also gives plant roots more room to grow. Without dung beetles, animal feces would reduce grazing land, and would provide ideal breeding habitat for disease-spreading flies and parasites, including worms and bacteria. These could greatly reduce animal populations and decrease biodiversity. Hardened cow feces can last years. Dung that is not decomposed can cover grasslands or pastures, and convert them from carbon sinks to carbon emitters. So dung beetles greatly reduce atmospheric carbon while conserving it. Also, since methane depletes oxygen, the reduction of the release of methane into the air by dung beetles and other life helps maintain higher atmospheric oxygen levels.

Methane is also kept out of the atmosphere nonbiologically by the formation of methane hydrates at low temperatures and high pressures in the seafloor in the deeper parts of continental shelves. These are frozen forms of methane, most of which was generated by methanogens. They are also frozen in the Arctic and Antarctic. If they are heated too much, the methane can escape and warm the Earth, potentially to temperatures catastrophic for biodiversity. If organisms did not sequester carbon through a long period of geologic time, the Earth would be so hot that the methane hydrates would melt, releasing huge quantities of methane into the air, making Earth too hot for most life, in a positive feedback loop.

Life evolved another way to conserve carbon while regulating temperature, which is the evolution of more efficient photosynthesis. Photosynthesis converts carbon dioxide into carbohydrate through a pathway with several chemical intermediates along the way, catalyzed by the enzyme Rubisco. In most photosynthetic organisms, the first compound created from carbon dioxide in the photosynthetic pathway has a backbone of three carbon atoms, so this is called C3 photosynthesis, and the plants are called C3 plants. But certain plants, such as tropical grasses, create a molecule with a backbone of four carbon atoms as the first chemical created from carbon dioxide in the pathway of photosynthesis. This process is called C4 photosynthesis, and the plants are called C4 plants. C4 plants evolved a revolutionary upgrade of photosynthesis well after photosynthesis originally evolved. The upgrade comes in the form of a solar-powered carbon dioxide pump that increases the concentration of carbon dioxide around Rubisco. The result is a concentration of carbon dioxide in the plant's cells that is 10 times that in the air, allowing Rubisco in C4 plants to convert carbon dioxide to carbohydrates with an incredible efficiency, much better than in C3 plants. Tropical grasses evolved this much more efficient photosynthetic system about 30 mya, although there is a good deal of uncertainty about this date. C4 grasses expanded tremendously throughout most of the world 6 to 8 mya, with C4 grasslands and savannas replacing forests in many areas. C4 photosynthesis thrived and spread, and today there are about 7,500 species of C4 plants, occupying one-fifth of the vegetated land surface of Earth, and accounting for 30% of the primary productivity (carbohydrates produced by photosynthesis) of the terrestrial biosphere (Lloyd and Farquhar, 1994), although they are only 3% of plant species. Most C4 plants are confined to tropical and subtropical climates, because they function best in bright sunlight at high temperatures. By far the majority are subtropical grasses. They dominate subtropical grasslands and savannas. The ascendancy of C4 plants is one of the most profound transitions of global ecosystems in the Earth's history. The evolution and spread of a photosynthetic system that uses carbon more efficiently conserves carbon and prolongs the time that there will be enough carbon for photosynthesis and life on Earth's land, in the sea, and freshwater to continue. Studies show that life facilitated and even largely caused the evolution of C4 plants (Beerling, 2007). Millions of years before C4 plants appeared, life—mainly large land plants—caused carbon dioxide levels to fall precipitously. Many were buried, and their roots and their symbiotic fungi caused weathering. Both of these occurrences caused carbon dioxide to drop to extremely low levels. The resultant low levels of carbon dioxide caused natural selection to favor plants that used carbon dioxide efficiently when carrying out photosynthesis. Hence, some plants evolved the more efficient C4 system of photosynthesis. It took C4 plants millions of years to evolve after low atmospheric carbon dioxide was the norm. It is unclear why.

Earth's atmosphere is about 78% nitrogen, 21% oxygen, about 0.04% carbon dioxide, and the rest trace gases. The high concentration of nitrogen causes nitrogen to collide often with carbon dioxide molecules (Li et al.,2009). This broadens the infrared (heat) absorption band of the carbon dioxide molecule via an effect of quantum physics, increasing the strength of carbon dioxide's warming of the Earth. Thus, a high concentration of nitrogen in the atmosphere increases the greenhouse warming effect of carbon dioxide. The actions of some bacteria remove nitrogen from the air and add it to the sea. Some of this nitrogen makes it to the seafloor. Some types of bacteria help bury nitrogen in sediments in the sea bottom, and this becomes hard rock over a long time period. This is the major mechanism of removing nitrogen from the atmosphere and sequestering it. Nitrogen is removed from the system whenever an organism is buried, since all organisms have nitrogen in their biomolecules. However, bacteria are likely the organisms that bury the most nitrogen, although this is not known for sure. We do know that a significant amount of nitrogen is sequestered by life. This weakens the greenhouse effect of carbon dioxide and cools Earth without losing carbon dioxide from the system. Li et al. (2009) state that, by this process, life can extend the time of oxygen-generating photosynthesis and hence higher life and high biodiversity at least 2.3 billion years into the future, more than doubling previous estimates. Thus, it appears that life regulated temperature and cooled the planet as solar output increased throughout life's history by burying nitrogen as well as carbon. Also, if life had not buried nitrogen throughout its history, there could be an excess of it. This would cause the same problems that humans cause today when they add too much nitrogen to ecosystems. This includes phytoplankton and bacteria growing in excessively large numbers due to the abundance of nitrogen, which is a nutrient they need. When they die, other bacteria decompose them, and this requires enormous amounts of oxygen. Thus, oxygen gets depleted, and animals from fish to crabs die off in great numbers. This process, called *eutrophication*, can occur in both freshwater and marine ecosystems. There are dead zones with essentially no higher life in several areas in the seas as a result of this today. These dead zones were caused by humans adding too much nitrogen to the oceans. A well-known dead zone in the Gulf of Mexico results from run-off of agricultural and other wastes and delivery of nitrogen loads into the Gulf of Mexico. Also, freshwater ecosystems also have die-offs from eutrophication due to excess nitrogen as a result of human activity.

The result of this conservation of carbon by life is that life's beneficial sequestering of carbon will not cause a catastrophe. Life is sequestering carbon and hence keeping the temperature and acidity in aquatic ecosystems favorably low for itself, but keeping the rate of carbon loss from the biosphere so slow that nonbiological factors will kill off life before life's burial of carbon removes so much carbon from the biosphere that life is significantly harmed. The catastrophe of the distant future will be caused by a nonbiological force, the increased heat output of the sun, not life removing carbon from the biosphere. As the sun continues to increase its heat output, the weathering rate of silicate minerals will increase. Recall that weathering sequesters carbon, so this will result in a decrease in the atmospheric carbon dioxide necessary for photosynthesis. As the increased heat causes water to evaporate from the Earth's surface, rocks will harden, causing plate tectonics to slow and eventually stop. This will cause all volcanoes to become inactive and stop adding carbon to the atmosphere and minerals to the soil. Water vapor is a greenhouse gas, so the Earth will get even hotter. And there will not be sufficient liquid water when it evaporates. In about 600 million years, the level of carbon dioxide will be too low for C3 plants to photosynthesize. Plants that use C4 photosynthesis can survive at carbon dioxide concentrations as low as 10 parts per million, but in time, carbon dioxide will be too low for them to carry out photosynthesis. All life on Earth's surface, including in fresh and salt water will suffer the same fate, for all life needs carbon. Thus, the sun's increased heat output will cause the biosphere to have too little carbon for any life to survive except possibly life that is deep underground, by shortly after 600 million years from now, a time period about 1.7 billion years less than the 2.3 billion years that Li at al. estimate biology is extending the time life will take to significantly deplete carbon. (Note: Increased temperature from the current climate crisis melts the permafrost, adding greenhouse gasses to the atmosphere, but at the higher temperatures discussed in this paragraph, atmospheric carbon dioxide is greatly *decreased*).

Life Likely Regulates Temperature with Negative Feedback and Dimethyl Sulfide

There is another important homeostatic, negative feedback process that regulates Earth's temperature, also driven by biology. This one is independent of carbon dioxide and methane. The CLAW Hypothesis, which takes its acronym from the first letters of the surnames of its four originators (Charlson, Lovelock, Andreae and Warren), proposes that particular phytoplankton that produce the chemical gas called dimethyl sulfide (DMS; chemical formula CH3-S-CH3, or C2H6S) are responsive to variations in temperature, and that these responses lead to a negative feedback loop that acts to stabilize the temperature of the Earth's atmosphere (Charlson et al., 1987). Here is how the negative feedback loop works. If the sun increases its output, phytoplankton increase their growth rates by a physiological response to higher temperature, and by increasing their photosynthetic rate due to enhanced solar radiance. Thus, there will be more phytoplankton. Some phytoplankton make a chemical called dimethylsulfoniopropionate (DMSP; chemical formula (CH3)2S+CH2CH2COO–, or C5H10O2S). Their increased growth increases their DMSP production. When these phytoplankton are damaged or killed, for example by grazing zooplankton, a virus bursting them open, or turbulence, their DMSP is released. Then bacteria and other phytoplankton convert the DMSP to DMS, a gas essential to Earth's biogeochemical cycles that easily leaves the ocean's waters. Some of it goes into the air. DMS is also released by some phytoplankton when they are grazed by herbivores, such as tiny fish, copepods, invertebrate larvae, and small protozoa called ciliates. DMS represents 95% of the sulfur that travels from the sea to the air, and supplies about 50% of global sulfur to the air from life. DMS is oxidized with the sun's aid to sulfur dioxide in the atmosphere, leading to the synthesis of sulfate aerosols, which attract molecules of water to form water droplets that make up clouds. This increases cloud droplet number, elevating cloud water content and cloud area. The effect of clouds on weather depends on the cloud type. Some clouds reflect heat exiting the Earth back to the ground, enhancing the greenhouse effect and increasing air temperature. However, the clouds produced by these sulfate aerosols increase the cloud reflectivity of radiation coming toward Earth from the sun, so the clouds reflect incoming sunlight, and thus lower the temperature. This is what most clouds do. About a third of solar radiation that reaches Earth's atmosphere is radiated back by clouds and surfaces like snow and ice. On the other hand, less solar input and heat from the sun leads to less DMS, less cloud cover, and an increase in atmospheric temperature. The CLAW hypothesis states that DMS is a mechanism by which life regulates and stabilizes the climate significantly by negative feedback on a large scale. Therefore, it is compatible and supportive of both the Gaia Hypothesis and the Pachamama Hypothesis. The effect is significant, since marine phytoplankton that synthesize DMSP cover a large area of the sea at any given time. Some seaweeds, including some red and some green algal species, also produce DMSP, and have the enzyme to convert it to DMS. How much they affect cloud formation is not known. There is empirical evidence supporting the CLAW Hypothesis (Andreae et al., 1995; Cropp et al., 2005), although it is equivocal (Vallina et al., 2007), and some have suggested that CLAW-like negative feedback may operate in the planet's sulfur cycle without an active biological component (Shaw et al., 1998).

Phytoplankton release DMS when they are eaten by krill. Savoca and Nevitt (2014) showed this is beneficial to the phytoplankton because far-ranging seabirds such as petrels and albatross use the smell of DMS as a cue to locate the prey they eat, which includes krill. The sea is so uniform that it is hard for seabirds to locate their prey by sight. The phytoplankton attract predators of the herbivores that eat them. These seabirds have excellent senses of smell. Seabirds also fertilize the sea, aiding plankton with iron from their feces, in this symbiosis between phytoplankton and seabirds. In the vast Southern Ocean, iron is a limiting resource that phytoplankton need. Thus, these predatory seabirds are tied to both global climate regulation and the health of the ocean ecosystem through a symbiotic relationship with phytoplankton. This symbiosis between phytoplankton and seabirds has also been shown between phytoplankton and various other animals that use the smell of DMS as a cue to find krill to eat, including penguins, seals, sharks, sea turtles, coral reef fishes, and possibly whales. Plastic that has been soaking in the sea emits DMS. This attracts some birds, some fish, and loggerhead sea turtles (*Caretta caretta*) (Pfaller et al., 2020). All of these animals then mistakenly eat the plastic and are harmed by it.

Viruses are also involved in this DMS system, since DMSO is released when they burst open and kill the cells of one-celled organisms such as bacteria and phytoplankton (Fuhrman, 1999). Thus, the DMS negative feedback loops that act as thermostats and regulate and stabilize the atmosphere's temperature involve viruses, phytoplankton, krill, red and green algae, fish, sea turtles, sea birds, penguins, and marine mammals. Therefore, all of these organisms help regulate temperature. In reviewing aspects of the influence of phytoplankton on climate, Watson and Liss (1998) speculated that the overall effect of phytoplankton today is to cool the planet by about 6°C (about 10.8°F), with one-third of this figure being due to carbon burial and two-thirds due to cooling by DMS. Regulating temperature by negative feedback using DMS is another way life controls temperature while conserving carbon in the biosphere.

In addition, a key process in the entire planetary sulfur cycle is the transfer of sulfur compounds from the ocean to the land via the atmosphere, and DMS is the dominant marine sulfur compound *generated by life* in this process. DMS is therefore essential to the global sulfur cycle and in transferring sulfur from the sea to the air and from there to the soil. DMS is transferred from the air to the soil by rain, and is easily incorporated into soil, elevating its sulfur content. So biology helps add sulfur to the soil from the sea. This is beneficial to soil life, and the process does not tend to add too much of this nutrient. Since sulfur compounds significantly alter temperatures, cloud formation, and hence rainfall, they also significantly influence the global water cycle. So the sulfur and water cycles influence each other via biology, and are both of crucial importance to life and profoundly influenced by it in feedback loops wherein life helps life.

Organisms Regulate Silicon, Helping Life

In addition to regulating carbon, organisms regulate silicon, which has similar chemical properties to carbon. All naturally occurring types of silicon are nontoxic and it is even required by life. But high concentrations of soluble silicon compounds in water may impede the use of the very important element, phosphorus, by organisms, and limit algal growth. Silicon is abundant in the ocean, with 80% coming from rivers. It originates from weathering, volcanoes under the sea, and sea sediments. Some of it had to be removed to bring its levels closer to what is optimal for life. Organisms played and still play a large role in this. Plankton ingest it and use it for various processes. This is a major mechanism of removal, with diatoms removing over 6.7 billion metric tons of silicon each year from the ocean (Treguer, 1995). Diatoms have cell walls made of silica (silicon dioxide), which contains silicon. Freshwater diatoms also remove silicon from rivers (Roubiex et al., 2008). Sponges remove large amounts for their skeletons. Various plant species remove it. Plants contain about 200–62,000 parts per million by dry weight of silicon. Dandelions and bamboo contain silicon in their stems and leaves, increasing their stability. In all these groups of organisms, silicon is sequestered by burial upon death. Yet, also, by storing silicon in their cell walls, skeletons, stems, and leaves, organisms keep a large quantity of silicon available to the biosphere, while preventing it from doing harm as a result of being too abundant. Silicon is also removed nonbiologically by reactions of dissolved silicon with clay minerals and by its settling into sediments such as the seafloor and muddy lake bottoms. But biology removes more silicon and hence is more important for silicon regulation than nonbiological processes.

There is a Good Chance that Biology Aids Life by Regulating the Ocean's Salt Content

Organisms might also contribute to regulating the salinity of the ocean, which has been constant at about 3.5% for a very long time (Segar, 2012). This is important for sea life because the cells of most organisms need a stable salinity, and cannot survive at levels above 5%. Sea organisms have adaptations to regulate the amount of salt they take in from seawater, but this ability is not unlimited and takes a good deal of energy. Salinity that is too high is detrimental to life. Salts from the weathering, erosion, and dissolution of rocks have been transported by rivers and deposited into the oceans through the ages, and this would increase the salinity in the sea if there were no counterbalancing force, and none is known. Salinity in the sea might be greatly reduced by seawater circulation through hot volcanic rocks at the bottom of the ocean, and its emergence from hot water vents on mid-ocean ridges. However, this filtration might not sufficiently regulate the sea's salt concentration, and seawater's salinity is far from what it would be if no outside forces were affecting it, allowing for the possibility that biology may have made it favorable to life. Segar (2012) hypothesized that bacterial colonies reduce the sea's salinity by removing salt from seawater and forming salt plains on the seafloor. If so, bacteria are key to keeping seawater from being too salty for organisms to thrive. More research is needed before we can reach any definitive conclusions.

Venus and Human Impacts Show that Organisms Have Profoundly Helped Life by Regulating Atmospheric Greenhouse Gases and Temperature

A null hypothesis is one that is the alternative to a proposed hypothesis. The null hypothesis with respect to the ABH/Pachamama Hypothesis is that life does nothing to increase biodiversity, help life, or make Earth better for life. The planet Venus serves as a good control to test the Pachamama Hypothesis against the null hypothesis, showing what happens when there is no life to sequester carbon from the atmosphere. Venus is closer to the sun than Earth is, but this is a small percentage of the cause of Venus being well over 800°F (about 427°C) hotter than Earth. In this control and test of the null hypothesis, there is no life to sequester the carbon, and hence nothing to prevent a runaway greenhouse effect. Outgassing of carbon dioxide and other gases from more than 1,000 volcanoes or volcanic centers larger than 12 miles in diameter on the surface of Venus has caused surface temperatures to reach 900°F (about 482°C). Hence, we can reject the null hypothesis that greenhouse gas levels would be favorably low for life without the action of biological organisms. On Earth, without the sequestering of carbon by life, there would be about 1,000 parts per million of carbon dioxide in the atmosphere, which is much less favorable to life than the pre-industrial levels.

One part of the null hypothesis is that if life did not sequester greenhouse gases such as carbon dioxide and methane, their higher levels would make no difference in diversity. Interestingly, but unfortunately, the effects of increased atmospheric carbon dioxide, methane, and other greenhouse gasses are currently being tested in an unintended experiment by the addition of these greenhouse gases to the atmosphere by humans. Preindustrial levels of atmospheric carbon dioxide were about 260 to 270 parts per million (Wigley, 1998). In early 2015, the seasonally adjusted concentration was about 400 parts per million, with a recent growth rate of between two and four parts per million per year. Humans have warmed the Earth by an average of 1.1°C (2°F) since the Industrial Revolution.

This chapter will discuss only effects on biodiversity, since effects exclusively on humanity are not relevant to this test of the Pachamama Hypothesis. Following are some of the major effects of human-induced increases in greenhouse gasses. Life can adapt to changes, but the changes listed below are generally too fast and extreme for much of life to adapt to.

The hotter water is, the less oxygen it can hold, and a rapid increase in water temperatures creates low oxygen content in both fresh water and ocean systems. This harms life in both these systems. Fish and invertebrate populations have been negatively impacted in rivers and streams as a result.

Melting of coastal ice in Greenland is causing more fresh water to enter the ocean there. The current system needs the water in this area to have high salt content so it can be dense enough to sink in order to power ocean currents, for it is the sinking that powers the currents. Without currents, there are tremendous temperature differences in different areas of the sea, and this is very bad for sea life. Many species also rely on currents for transport. Also, the currents interact with the atmosphere. Thus, if the Gulf Stream loses strength, it will not warm Europe, and Europe will becoming extremely cold, which will be highly detrimental to life.

Climate change is causing animals that live in mountains to move to higher altitudes, and others to move toward the poles. This is disrupting ecosystems, causing species that are not adapted to each other to live together, thus decreasing population sizes, sometimes to zero. Some species are running out of space to move higher on the mountains they inhabit.

Animals and plants are changing the timing of their life cycles. Species that depend on other species are becoming out of synchronization with them. Insects that eat nectar or pollen and pollinate flowers are finding their host flowers are no longer blooming when the adult insect populations are peaking. Thus, the insects have too little food and the plants cannot reproduce as effectively because the insects do not pollinate their flowers. Birds that eat insects are no longer nesting when insect populations peek, so are having a hard time feeding their young. This lack of synchronicity of flowers, insects, and birds is due to the fact that these groups respond to climate change differently. It is causing large declines in several species in all of these groups, and will likely result in the extinctions of some of these species.

Bears in the United States are emerging from hibernation earlier, making it harder for them to obtain food and making them more prone to starvation.

Life's regulation of climate has kept moisture levels unfavorably low for some fungal species in many areas, and human-induced increased temperatures have increased moisture levels in many of these areas, creating favorable conditions for certain fungal pathogens. Many suspect such climatic change played a role in the following fungal diseases of animal species, because they only became serious epidemics after climate change increased humidity in the disease areas. Chytrid fungi cause a disease called chytridiomycosis in amphibians that has caused dramatic population declines and even extinctions of amphibian species. As of 2012, white-nose syndrome (WNS), caused by the fungus *Pseudogymnoascus destructans*, was estimated to have killed 5.7 million to 6.7 million bats in North America (Froschauer and Coleman, 2012). As of 2017, 15 bat species, had been impacted by WNS. The often-lethal fungus, *Ophidiomyces ophiodiicola*, has infected 23 unrelated species of snakes from the eastern to midwestern United States. Since white band disease was first reported in the 1970s, the disease has led to the devastation of approximately 95% of the elkhorn coral and staghorn coral in the Caribbean. This is not caused by a fungus, but a bacterium. Warmer water causes the bacteria to shift from a species of nonpathogenic bacteria being dominant to a disease-causing species of bacteria of the genus *Vibrio* becoming dominant.

Since the 1970s, droughts have become more frequent, longer, more intense, and have covered larger areas, notably in the tropics and subtropics. Higher temperatures allow air to hold more moisture, so it does not fall as rain. This causes droughts in the tropics and subtropics, which reduces the number of trees and diversity of tree species in these ecosystems, and it weakens those that manage to live. Drought increases the number, intensity, and size of fires in rainforests, making the amount of fire far above the optimum for life. There were no large fires in rainforests before climate change. Since human-caused climate disruption started, there have been major fires in rainforests in the Amazon, Mexico, the Congo Basin, and Southeast Asia.

Drought is also causing increased fires and insect outbreaks in temperate forests. Partly as a result of drought, fires are becoming more frequent and larger in the western United States. This releases more carbon dioxide into the air in a positive feedback loop. Between 1997 and 2010, over 5 million hectares (19,305 square miles) were affected by bark beetles in the western United States, mainly the mountain pine beetle, spruce beetle, and pinyon ips beetle. These insect attacks release tremendous quantities of carbon dioxide in another destabilizing positive feedback loop. In many areas, a threshold has been crossed, and forests will not grow back after being burned, because of the severity of drought. Conditions in some forests have become too dry or hot to support them.

Life Regulates the Atmosphere's Greenhouse Gas Levels and the Earth's Temperature

In the Amazon, a forest mortality threshold is looming, and a good deal of rainforest could be permanently converted to savanna. Around the world, research has suggested that the tree mortality rate in some temperate and tropical forests has doubled or more in recent decades. As forests die back, they switch from storing carbon to emitting it, in a maladaptive positive feedback loop.

Coral reefs, the most diverse marine ecosystems on Earth, are dying due to coral bleaching, a process whereby the symbiotic algae that live in the coral leave the coral when the sea's temperature rises beyond a threshold temperature. The algae carry out photosynthesis, providing carbohydrate to themselves and the animal part of the coral. Without the algae, the coral dies and the fish that depend on the coral die. There is a tremendous loss of species.

Upwelling is the movement of water that is rich in nutrients from deeper water in the ocean to surface waters, bringing nutrients to phytoplankton, powering the major food webs of the sea. When temperatures rise, upwelling occurs less often or ceases completely. Entire food webs can crash.

Warmer winters will cause many reptile species to decrease in numbers of individuals or perish because they need cold to lower their metabolism and prevent starvation during hibernation. The sex of crocodiles, alligators, and turtles is determined by the temperature the eggs are incubated at. If the temperature is raised even a few degrees, most or all offspring would be the same sex, causing population declines or extinctions.

Glaciers in mountains supply rivers with a major portion of their water all over the world. As the planet gets hotter, glaciers worldwide are melting and not being replenished as fast as they are being lost. When the glaciers are melted, there will be no source of water for rivers, and they will dry up. Melting of the Himalayan glaciers is removing the water supply for seven rivers in Asia that support many freshwater species. Drying out of the rivers will result in the total destruction of riparian, wetland, forest, and other habitats, and the loss of essentially all the species in all of these habitats. Entire forests will die from desiccation. This is a huge area, including Tibet, Nepal, China, India, Pakistan, Bangladesh, Cambodia, Laos, Myanmar (Burma), Thailand, Vietnam, and more. The Earth is also losing glaciers in the Sierra Nevada, which supply the water for major rivers in California; the Rocky Mountains, which supply the Colorado River; the Andes, which supply the Amazon River system; the Alps; and other mountain ranges. This will result in the drying of all the rivers these glaciers supply water to. The Amazon rainforest, the most diverse terrestrial ecosystem on Earth, will have very little or no water, resulting in the extinction of millions of species of animals and plants.

The Amazon is the world's largest tropical rainforest. It has played a key role in absorbing and storing carbon. Deforestation has had a great impact on it. Gatti et al. (2023) found that the western Amazon now is a net carbon source, and no longer a sink. It released about 20% more carbon dioxide into the atmosphere than it took in from 2010–2019. The Amazon's ability to act as a major carbon sink has declined as a result of deforestation and climate change. Increased temperatures are promoting changes in dry season conditions and stress to trees, causing them to emit more carbon. The authors found that an increase in deforestation and intensification of the dry season that results from this and global climate change promote increases in fire occurrence, ecosystem stress, and carbon emissions in the eastern Amazon.

Drying of cloud forests has driven the golden toad (*Incilius periglenes*) of Costa Rica extinct. Warming has made conditions drier in the cloud forest of Monteverde National Park in Costa Rica, where the average altitude of clouds has been rising since the 1970s. This caused this toad to die off because it needs moist conditions. Reptiles, amphibians, and other types of animals are dying from the drying of cloud forest.

Many polar species are losing habitats. Polar bears, seals, penguins, and other Arctic and Antarctic wildlife need sea ice and are experiencing population declines as their sea ice habitat decreases.

Methane hydrates are found in certain areas in the sea, some lakes, and in the Arctic and Antarctic permafrost. They consist of the powerful greenhouse gas methane under low temperatures, forming a solid similar to ice. If the Earth were heated to high enough temperatures by greenhouse gases, methane hydrates would melt and release tremendously high quantities of methane into the air, heating the Earth yet more, which would cause more methane hydrates to melt and release more methane into the atmosphere, in a devastating positive feedback loop, making Earth extremely hot and unfavorable for life, causing a great decrease in diversity, even a probable mass extinction. The Permian-Triassic mass extinction of about 252 mya, the largest extinction in Earth's history, probably resulted from a rise in temperature from volcanoes releasing greenhouse gasses that led to the melting of methane hydrates, and thus the release of large quantities of methane from methane hydrates. The resulting extreme temperature increase was likely the main factor in this mass extinction. Methane hydrates contain as much as 20% of all carbon on Earth. All the living organisms on the planet together have less carbon in them than there is in Earth's methane hydrates.

Melting of the Arctic and Antarctic glaciers will cause a large rise in sea level, destroying coastal habitats, such as mangroves and other wetlands, with a devastating effect of biodiversity.

A report from the UN's Intergovernmental Panel on Climate Change warned that human-induced global warming is threatening large numbers of species with extinction, even under relatively conservative estimates (O'Neill et al., 2017). If Earth warms by 1.5°C (2.7°F), up to 14% of all plants and animals on land will face a high risk of extinction. This temperature rise is almost certain to occur. With a 3°C (5.4°F) temperature rise, up to 29% of species on land could face extinction. A shocking 47% of species have already lost some of their populations because of climate change. Warming of only 1.5° C has the potential kill up to 90% of tropical coral reefs, which support over 25% of all marine species.

The above list of ways that climate change from greenhouse gases can devastate biodiversity, and indeed is doing so, is not comprehensive. It shows that we can reject the null hypothesis that the sequestering of carbon by life has no effect or a negligible effect on biodiversity.

The Detrimental Impacts of High Atmospheric Carbon Dioxide Levels on Life Are Not Limited to Temperature Effects

There are several extremely severe negative impacts on biodiversity that high carbon dioxide levels produce, quite apart from temperature effects. The higher the atmospheric concentration of carbon dioxide, the greater the amount of it absorbed by the oceans. This is in fact happening today, and it acts to alleviate global warming to some extent. However, the excess carbon dioxide in the seas reacts with water to form carbonic acid, raising the acidity of the oceans. The higher the acidity of water, the less calcium can come out of it as a solid, so more of it stays in solution. The less calcium is able to come out of solution as a solid, the less organisms can use it to form their protective calcium shells. Groups of organisms with protective calcium shells or skeletons include myriad species of: of phytoplankton; bivalve mollusks, such as clams and mussels; sea snails; barnacles; crustaceans from crabs to shrimp to the tiny krill that feed many species of great whales and several other species; sea urchins; sand dollars; and coral. Thus, if organisms did not sequester carbon, acidity in the seas would cause great decreases or extinctions of a number of species in all of these animal groups. This would severely impact all groups that depend on these animals, directly or indirectly. Entire food webs would be disrupted. Any species directly or indirectly dependent on coral reefs for food or shelter would decline or become extinct. This includes the approximately 25% of all marine fish species that live in coral reefs. And the high acidity would also disrupt the reproduction of many species, alter biogeochemical cycles, disrupt other physiological processes of marine organisms, and damage marine ecosystems in other ways. Increased temperature and acidity would likely act together synergistically, with tremendously negative effects on ecosystems (see Doney, 2006; Fabry et al., 2008). Animals that build shells and live in freshwater rivers, ponds, and lakes would also be impacted by increased acidity of their water. This includes freshwater clams, mussels, snails, shrimp, and so on. So *increased carbon dioxide in aquatic ecosystems has the potentially to cause absolutely huge decreases in diversity.* This effect has already started to manifest itself in several ecosystems.

Life Regulates the Atmosphere's Greenhouse
Gas Levels and the Earth's Temperature

A study by Zhili et al. (2012) of soil microbes found high levels of carbon dioxide in soil causes lower bacteria population numbers and lower numbers of bacterial species. Reduced microbial diversity would reduce diversity of many eukaryote species, including soil algae, earthworms, plants, and other groups, since bacteria are a food source and carry out processes vital to these organisms. The reduction of soil algae, earthworms, plants, and other groups would cause the decline of the myriad species that benefit from them. Bacteria have high populations and short generation times, so may be able to adapt to high carbon dioxide, at least to some extent.

Plants that grow in very high atmospheric carbon dioxide grow faster, but have a low protein to carbohydrate ratio. In general, insect larvae feeding on plants grown in high carbon dioxide conditions eat more in response to the lack of sufficient protein, but still become protein deficient. They are more subject to predation, malnutrition, starvation, and overall death rates, and smaller if and when they do emerge as adults. Adult spittlebugs, which suck sap from plants, had their survival rates decrease 27% in high carbon dioxide (Brooks and Whittaker, 2001). In contrast, aphids reproduce 10 to 15% faster in elevated carbon dioxide conditions. Infested bean plants grown in high carbon dioxide were so overwhelmed by the aphids' rapid population growth that they could not grow flowers or new shoots (Awmack et al., 2003).

A decrease of insect herbivores would have serious negative effects on ecosystems, because most insect species are herbivorous, and there are about 900,000 known species of insects, which is about 80% of all animal species. Most insect species have not been described; it is estimated there are a total of 20 to 30 million insect species, and there are about 1019 insects alive at any given time. If there were too few plant-eating insects, the plants they eat would not be regulated, so they would become so dense that many animal species would no longer be able to live in them, and they would deplete soil nutrients, leading to the death of great numbers of plants, and the animals that eat them. Also, plants would crowd each other and compete for space, water, and nutrients. The poorer competitors would die off, so there would be a much lower number of species of plants. A decrease in herbivorous insects would also tremendously negatively impact all species they aid, such as their predators, the plants they pollinate, and plants whose seeds they disperse.

Herbivorous mammals such as deer also suffer from lower protein if they consume plants grown in high carbon dioxide. Their loss would negatively impact the plants they eat, their predators, and many other species that they benefit. The fact that plants grown in high carbon dioxide have a low protein-to-carbohydrate ratio has the potential to cause tremendous reductions in diversity in ecosystems, even causing them to crash.

Did Biology Create Nearly Optimal Conditions for Life?

None of the following was done intentionally by life. As stated in Chapter One, the Pachama Hypothesis is naturalistic and scientific, and does not rely on teleology. Methanogens kept the early Earth warm when the sun was weak. Life continuously removed greenhouse gases from the air as the sun increased its heat output through time. Thus, life regulated greenhouse gasses and the temperature, keeping Earth in a favorable, cool temperature range. It is possible that organisms come close to optimizing the levels of oxygen, water vapor, carbon dioxide, and methane in the atmosphere for life. Biology seems to regulate the two major greenhouse gases, carbon dioxide and methane, by negative feedback, keeping them at favorable levels to create temperatures beneficial to life. Organisms seem to regulate the amount of oxygen in the atmosphere, keeping it at near-optimal levels for life. Life increased the atmospheric oxygen level from less than 1% to today's life-favorable 21% by producing oxygen through photosynthesis and burying reduced carbon. (Recall that reduced carbon is carbon that is not combined with oxygen). Because of this, when it is buried, there is a net gain of oxygen in the air. This contrasts with carbon that is combined with oxygen, the burial of which does not necessarily result in an increase in atmospheric oxygen. The current atmosphere is 21% oxygen due to the actions of organisms. This is highly beneficial to life. Indeed, it is needed for higher life to exist. Oxygen levels are sufficient to provide the energy needed for higher organisms. But at excessive levels, oxygen would be harmful to cells and create damaging chemicals called free radicals. In regulating the level of atmospheric oxygen, biology may also come close to optimizing the amount and intensity of fires for life to thrive. In regulating the amount of carbon dioxide and methane in the air, life influences the temperature and hence the amount of water vapor in the air. The higher the temperature, the more water evaporates and rises into the air. The amount of water vapor influences the amount of cloud formation and hence the amount of lightning, which is generated by clouds. The amount of lightning influences the amount of fire. Thus, life may cause a favorable amount of water vapor in the air to help cause a life-favorable amount of fire. The atmospheric oxygen level is also important in determining the number and intensity of fires. The more oxygen there is in the air, the more easily fires burn, because fire uses oxygen as a fuel to burn. If life did not produce the intermediate atmospheric oxygen levels we see today, there would not be the favorable, intermediate amount of fire for life to thrive. Life may even keep the atmospheric oxygen level at a biology-favorable 21% by negative feedback (see Chapter 4). This also optimizes the amount and intensity of fires. Too much fire is bad for forests, but so is too little. Forests need fires to recycle nutrients and create fertile soil (see Chapter 4). Life produces almost all of the oxygen and most of the methane, which combine with and destroy each other. Given its depletion by oxygen, to sustain the amount of methane in the Earth's atmosphere, at least a trillion kilograms (a billion tons) of it must be added to the air annually. Most of this is made by life. It was needed to maintain a warm enough temperature before humans started heating the planet. Twice as much oxygen as methane must be added back to the air because methane destroys it, and this addition of oxygen is almost entirely done by life. Life produces just enough of both oxygen and methane that neither depletes the other or becomes too abundant. Also, the level of ozone in the ozone layer might be close to an optimal balance, with enough ozone to protect life from excessive UV radiation but a small enough amount of it to admit enough UV to induce a favorable level of mutations for evolution and adaptation. This ozone level is due to the level of oxygen created mainly by life.

It is also possible that the actions of organisms resulted in greenhouse gas levels nearly optimal for the best number and intensity of hurricanes for the intermediate disturbance of coral reefs, maximizing their diversity. Higher temperatures cause conditions for more hurricanes, and more intense ones.

Too many hurricanes and ones that are too intense would be destructive to coral reefs, but too few would allow the better competitors among the coral species to drive less competitive species locally extinct, reducing the number of species of coral and hence of fish and invertebrate species dependent on coral. An intermediate number and intensity of hurricanes results in the highest biodiversity on coral reefs.

These ideas that life creates optimal levels of greenhouse gases, oxygen, water vapor, temperatures, fire, and ozone for life are hypotheses I am proposing that need testing. They are not meant to be accepted as established truth.

Interestingly, Earth's temperature and carbon dioxide levels have remained remarkably stable and close to optimal for life from about 10,000 years ago to just before industrial times. There have been periods when temperature departed considerably from the mean, such as the Little Ice Age (not a real glacial period; it was a period of lower temperatures that lasted from about the 16th to the 19th centuries), but these have been small fluctuations compared to some of the large fluctuations throughout the last 500 million years. This raises the possibility that life brought global temperatures and greenhouse gas levels to near-optimal levels by about 10,000 years ago, and kept them there, until humans intervened. It is striking how preindustrial atmospheric greenhouse gas levels were so close to optimal for life. These levels appear largely created by life. The amount that these life-favorable greenhouse gas levels aided life is spectacular. At these levels, droughts and floods were minimized, glaciers did not disappear and kept replenishing rivers with water, rainforests did not suffer droughts and major fires, forests of all types had intermediate and diversity-maximizing levels of fire and moisture and insect attacks, coral reefs thrived in favorable water temperatures, the size and frequency of hurricanes were near-optimal for coral reef diversity, trees caused life-favorable amounts of local rainfall and atmospheric moisture, the sea and freshwater ecosystems were at beneficial levels of acidity for life, there were near-optimal carbohydrate-to-protein ratios in plants for herbivores, ocean currents were favorable to life, there were near-optimal temperatures for ocean upwelling and hence phytoplankton blooms and healthy ocean food webs and healthy kelp forests, and many more favorable conditions for biology. One may wonder if life evolved and adapted to get better at controlling greenhouse gas levels and temperature until it was extremely good at it starting 10,000 years ago (not consciously, of course). One may also wonder if life evolved and adapted to get better at regulating levels of oxygen, ozone, and fire.

But we must be aware that the continents are approximately at their maximum possible distances apart, which may be optimal for life by allowing favorable sea currents, winds, phytoplankton blooms, and temperatures. The continents were clumped together as one land mass more than once in the geologic past, and will in time move back together. When the continents are together as one continent, oceans currents are different and probably less favorable for keeping temperature differences between different areas in the sea low. Today's currents make temperatures in the sea more uniform from one place to another, which is good for life. If the continents were currently clumped together, perhaps the current greenhouse gas levels would not be so close to optimal for currents. We do not know what greenhouse gas levels and temperatures will be when the continents coalesce again. Nor do we know to what extent life and perhaps nonbiological forces caused the highly life-favorable greenhouse gas levels and temperatures for the last 10,000 years. It is clear life played a large role, but we do not know precisely how large. The ABH, or Pachamama Hypothesis, is carrying out one of the functions of a hypothesis, which is revealing potentially fruitful areas of research. This question of to what extent life created nearly optimal or even approximately optimal conditions for life for various parameters such as atmospheric gasses and temperature is a particularly important and interesting area of research that should be pursued.

Conclusion

The sun is 30% hotter today than when life began. Life facilitated the warming of the young Earth by producing a blanket of atmospheric methane when the sun was cooler, then later cooled it by sequestering carbon, to life's benefit. Biology has also regulated other aspects of the physical-chemical environment to the benefit of life, and it still does. If organisms did not sequester carbon to the extent that they have, it is highly probable that the Earth would be so hot and aquatic ecosystems so acidic that only simple ecosystems with low biodiversity would exist on Earth's surface and in aquatic ecosystems.

References

Zeebe, R. E. & Caldeira, K. (2008). Close mass balance of long-term carbon fluxes from ice-core CO2 and ocean chemistry records. *Nature Geoscience* 1: 312-5.

Lovelock, J. E. & Whitfield, M. (8 April 1982). The lifespan of the biosphere. *Nature* 296: 561-3.

Lenton, T. & Watson, A. (2011). *Revolutions that Made the Earth.* Oxford Univ. Press, Oxford, U. K.

Lovelock (1979). *Gaia. A New Look at Life on Earth.* Oxford University Press, Oxford, New York, Toronto, Melbourne.

Frings, P. J. (Aug. 2019). Palaeoweathering: How do weathering rates vary with climate? *Elements* 15: 259–65. doi: 10.2138/gselements.15.4.259

Ueno, Y., et al. (23 March 2006). Evidence from fluid inclusions for microbial methanogenesis in the early Archaean era. *Nature* 440: 516–9. doi: 10.1038/nature04584

Haqq-Misra, J. D., Domagal-Goldman, S. D., Kasting, P. J., & Kasting, J. F. (3 Feb., 2009). A revised, hazy methane greenhouse for the Archean Earth. *Astrobiol.* 8 (6): 1127–37. https://doi.org/10.1089/ast.2007.0197

Berner, R. A. & Kothavala, Z. (2001). Geocarb III: A revised model of atmospheric CO2 over phanerozoic time. *American Journ. of Science* 301 (2): 182-204. ajsonline.org

Pavlov, A. A., et al. (5 May 2000). Greenhouse warming by CH4 in the atmosphere of early Earth. *Journ. of Geophys. Research; Planets* 105 (E5): 11981-90. https://doi.org/10.1029/1999JE001134

Morse, J. W. & Mackenzie, F. T. (1990). *Geochemistry of Sedimentary Carbonates.* Elsevier, Amsterdam, the Netherlands.

Miller, C. B. (2004). *Biological Oceanography*, pp. 94-5, 266-7. Blackwell Science Ltd., Oxford, U. K.

Lazier, J. R. N. (2006). *Dynamics of Marine Ecosystems*, p. 35. Wiley-Blackwell, Hoboken, N. J.

Boyd, P.W., Claustre, H., Levy, M., et al. (2019). Multi-faceted particle pumps drive carbon sequestration in the ocean. *Nature* 568: 327–35. https://doi.org/10.1038/s41586-019-1098-2

Danovaro, R., Corinaldesi, C., Dell'anno, A., Fuhrman, J. A., Middelburg, J. J., Noble, R. T., & Suttle, C. A. (Nov. 2011). Marine viruses and global climate change. *FEMS Microbiol. Rev.* 35 (6): 993–1034. https://doi.org/10.1111/j.1574-6976.2010.00258.x

Fuhrman, J. A. (10 June 1999). Marine viruses and their biogeochemical and ecological effects. *Nature* 399 (6736): 541-8.

Weinbauer, M.G. (2004). Ecology of prokaryotic viruses. *FEMS Microbiology Reviews* 28: 127–81.

Suttle, C. A. (15 Sept. 2005). Viruses in the sea. *Nature* 437: 356-61. doi: 10.1038/nature04160.

Proctor, L. M. & Fuhrman, J. A. (1990). Viral mortality of marine bacteria and cyanobacteria. *Nature* 343: 60–2.

Mojica, K. D. A. & Brussaard, C. P. D. (2015). The viral shunt in a stratified Northeast Atlantic Ocean. In Mojica, K. D. A., *Viral Lysis of Marine Microbes in Relation to Vertical Stratification*. UvA-DARE (Digital Academic Repository); Ch. 7. Univ. of Amsterdam, Amsterdam, the Netherlands.

Wilhelm, S. W. & Suttle, C. A. (Oct. 1999). Viruses and nutrient cycles in the sea: viruses play critical roles in the structure and function of aquatic food webs. *Bioscience* 49:781-8. https://doi.org/10.2307/1313569

Suttle, C. A. (Oct. 2007). Marine viruses—major players in the global ecosystem. *Nature Reviews Microbiol.* 5 (10): 801-12. doi: 10.1038/nrmicro1750. PMID 17853907

Blanc Mathieu, R., et al. (22 July 2019). Viruses of the eukaryotic plankton are predicted to increase carbon export efficiency in the global sunlit ocean. *BioRxiv.* doi: https://doi.org/10.1101/710228

Lawrence, J. E. & Suttle, C. A. (2004). Effect of viral infection on sinking rates of Heterosigma akashiwo and its implications for bloom termination. *Aquat. Microb. Ecol.* 37: 1-7.

Lawrence, J. E., Chan, A. M. & Suttle, C. A. (March 2002). Viruses causing lysis of the toxic bloom-forming alga, Heterosigma akashiwo (Raphidophyceae), are widespread in coastal sediments of British Columbia, Canada. *Limnol. Oceanogr.* 47 (2): 545-50. https://doi.org/10.4319/lo.2002.47.2.0545

Poorvin, L., Rinta-Kanto, J. M., Hutchins, D. A., & Wilhelm, S. W.

(2004). Viral release of iron and its bioavailability to marine plankton. *Limnol. Oceanogr.* 49: 1734–41.

Pachiadaki, M. G., et al. (24 Nov. 2017). Major role of nitrite-oxidizing bacteria in dark ocean carbon fixation. *Science* 358 (6366): 1046-51. doi: 10.1126/science.aan8260

Baumert, H. Z. & Petzoldt, T. (22 Oct. 2008). The role of temperature, cellular quota and nutrient concentrations for photosynthesis, growth and light-dark acclimation in phytoplankton. *Limnologica* 38 (3-4): 313–26.

Simon, N., et al. (Feb.-Mar. 2009). Diversity and evolution of marine phytoplankton. *Comptes Rendus Biologies* 332 (2-3): 159–70.

Siegel, D. A., et al. (8 March 2016). Prediction of the export and fate of global ocean net primary production: The EXPORTS Science Plan. *Frontiers in Marine Science*. Ocean Observation. doi: 10.3389/fmars.2016.00022

Watson, A. J. & Liss, P. S. (29 Jan. 1998). Marine biological controls on climate via the carbon and sulphur geochemical cycles. *Philosophical Transactions of the Royal Society B: Biological Sciences* 353 (1365). https://doi.org/10.1098/rstb.1998.0189. Print ISSN:0962-8436. Online ISSN:1471-2970

Mann, D. G. & Droop, S. J. M. (1996). Biodiversity, biogeography and conservation of diatoms. Biogeography of freshwater algae. *Hydrobiologia* 336: 19-32.

Fensome, R. A., MacRae, R. A., Moldowan, J. M., Taylor, F. J. R., & Williams, J. L. (1996). The early Mesozoic radiation of dinoflagellates. *Paleobiol.* 22: 329-38.

Prothero, D. R. (2004). *Bringing Fossils to Life: An Introduction to Paleobiology* (2nd ed.), pp. 210–13. McGraw Hill, Boston, MA.

Iglesias-Rodriguez, M. D. Halloran, P. R., Rickaby, R. E. M., Hall, I. R., et al. (2008). Phytoplankton calcification in a high-CO_2 world. *Science* 320 (5874): 336–40.

Marsh, M.E. (2003). Regulation of $CaCO_3$ formation in coccolithophores. *Comparative Biochemistry and Physiology Part B: Biochemistry and Molecular Biology* 136 (4): 743–54. doi:10.1016/s1096-4959(03)00180-5. PMID 14662299

Milliman, J. D. (1974). *Marine Carbonates*. Springer-Verlag, New York, N. Y.

Mitra, A., et al. (April 2016). Defining planktonic protist functional groups on mechanisms for energy and nutrient acquisition: Incorporation of diverse mixotrophic strategies. *Protists* 167: 106-20. http://dx.doi.org/10.1016/j.profits.2016.01.003

Mitra, A. (April 2018). Tiny creatures, part plant and part animal, may control the fate of the planet. *Scientific Amer.* https://www.scientificamerican.com/article/tiny-creatures-part-plant-and-partanimal-may-control-the-fate-of-the-planet/. (secondary literature).

Flynn, K. J. & Mitra, A. (1 Sept. 2009). Building the "perfect beast": Modeling mixotrophic plankton. *Journ. of Plankton Research* 31 (9): 965-92.

Stoecker, D. K., et al. (Jan. 2017). Mixotrophy in the marine plankton. *Annual Rev. of Marine Science* 9: 311-35.

Gibbs, S. J., et al. (30 Oct. 2020). Algal plankton turn to hunting to survive and recover from end-Cretaceous impact darkness. *Science Advances* 6 (44). doi: 10.1126/sciadv.abc9123

Jónasdóttir, S. H., et al. (29 Sept. 2015). Seasonal copepod lipid pump promotes carbon sequestration in the deep North Atlantic. *PNAS USA* 112 (39): 12122-6, doi: 10.1073/pnas.1512110112

Madin, L. P., et al. (May 2006). Periodic swarms of the salp Salpa aspera in the slope water off the NE United States: Biovolume, vertical migration, grazing, and vertical flux. *Deep Sea Research Part I: Oceanographic Research Papers* 53 (5): 804-19. https://doi.org/10.1016/j.dsr.2005.12.01

Perissinotto, R. & Pakhomov, E. A. (1998). Contribution of salps to carbon flux of marginal ice zone of the Lazarev Sea, southern ocean. *Marine Biol.* 131: 25-32.

Katija, K., et al. (3 May 2017). New technology reveals the role of giant larvaceans in oceanic carbon cycling. *Science Advances* 3 (5). e1602374. doi: 10.1126/sciadv.1602374

Ware, J. R., Smith, S. V., & Reaka-Kudla, M. L. (Sept. 1992). Coral reefs: sources or sinks of atmospheric CO2? *Coral Reefs* 11 (3): 127-30.

Kinsey, D. W., Hopley, D. W., Hopley, D. (March 1991). The significance of coral reefs as global carbon sinks— response to Greenhouse. *Palaeogeography, Palaeoclimatology, Palaeoecology* 89 (4): 363-77. https://doi.org/10.1016/0031-0182(91)90172-N

Lebrato, M., Iglesias-Rodriguez, D., Feely, R. A., Greeley, et el. (2010). Global contribution of echinoderms to the marine carbon cycle: a reassessment of the oceanic CaCO3 budget and the benthic compartments. *Ecol. Monogr.* 80 (3): 441–67. doi: 10.1890/09-0553. www.esajournals.org/doi/abs/10.1890/09-0553

Wilmers, C. C., et al. (2012). Do trophic cascades affect the storage and flux of atmospheric carbon? An analysis of sea otters and kelp forests. *Frontiers in Ecol. and the Environment* 10: 409–15.

Smith, C. (2005). Bigger is better: The role of whales as detritus in marine ecosystems. *Whales, Whaling, and Ocean Ecosystems* 12 (3): 1-46.

Higgs, N. D., et al. (1 June 2012). Evidence of Osedax worm borings in Pliocene (3 Ma) whale bone from the Mediterranean. *Historical Biol.* 24 (3): 269–77. doi:10.1080/08912963.2011.621167. ISSN 0891-2963. S2CID 85170976

Smith, C. R. & Baco, A. R. (2003). Ecology of whale falls at the deep-sea floor. *Oceanography and Marine Biology: an Annual Review* 41: 311–54.

Roman, J. & McCarthy, J.J. (11 Oct. 2010). The whale pump: Marine mammals enhance primary productivity in a coastal basin. *PLoS ONE* 5 (10): e13255. doi: 10.1371/journal.pone.0013255

Nicol, S., et al. (June 2011). Southern Ocean iron fertilization by baleen whales and Antarctic krill. *Fish and Fisheries* 11 (2): 203–9. doi: 10.1111/j.1467-2979.2010.00356.x

McInherney, F.A.; & Wing, S. (2011). A perturbation of carbon cycle, climate, and biosphere with implications for the future. *Ann. Review of Earth and Planetary Sciences* 39: 489–516. Bibcode:2011AREPS..39..489M. doi:10.1146/annurev-earth-040610-133431

Bowen et al. (2015). Two massive, rapid releases of carbon during the onset of the Palaeocene-Eocene thermal maximum. *Nature* 8 (1): 44-47. doi:10.1038/ngeo2316

Waddell, L. M. & Moore, T.C. (2006). Salinity of the Early and Middle Eocene Arctic Ocean from oxygen isotope analysis of fish bone carbonate. *Amer. Geophys. Union, Fall Meeting 2006, abstract# OS53B-1097.* http://www.agu.org/cgi-bin/wais?hh=OS53B-1097

Brinkhuis, H. & Schouten, S. (2006). Episodic fresh surface waters in the Eocene Arctic Ocean. *Nature* 441 (7093): 606-9. doi: 10.1038/nature04692

Pearson, P.N. & Palmer, M.R. (2000). Atmospheric carbon dioxide concentrations over the past 60 million years. *Nature* 406 (6797): 695–9.

doi:10.1038/35021000.http://www.nature.com/nature/journal/v406/
n6797/pdf/406695a0.pdf

Subramaniam, A., et al. (29 July 2008). Amazon River enhances diazotro-
phy and carbon sequestration in the tropical North Atlantic Ocean. *PNAS
USA* 105 (30): 10460-5. https://doi.org/10.1073/pnas.0710279105

Berzaghi, F., et al. (2019). Carbon stocks in central African forests enhanced by elephant disturbance. *Nature Geoscience* 12: 725–9.

Harris, N.L., Gibbs, D.A., Baccini, A. et al. (2021). Global maps of twen-
ty-first century forest carbon fluxes. *Nat. Clim. Chang.* 11: 234–40. https://
doi.org/10.1038/s41558-020-00976-6

Berner, R. A. (25 April 1997). The rise of plants and their effect on weathering and atmospheric CO2. *Science* 276 (5312): 544-6. doi: 10.1126/sci-
ence.276.5312.544

Berner, R. A. (Aug. 1992). Weathering, plants, and the long-term carbon
cycle. *Geochimica et Cosmochimica Acta* 56 (8): 3225-31.

Berner, E. K., Berner, R. A., & Moulton, K. L. (2003). Plants and mineral
weathering: present and past. *Treatise in Geochem.* 5: 169-88.

Quirk, J., et al. (22 Aug. 2015). Constraining the role of early land plants
in Palaeozoic weathering and global cooling. *Proc. of the Royal Soc. B. Biol.
Sciences* 282 (1813). doi: 10.1098/rspb.2015.1115

Lucas, Y. (2001). The role of plants in controlling rates and products
of weathering: importance of biological pumping. *Ann. Rev. of Earth and
Planetary Sciences* 29: 135-63.

Berner, R. A. (1998). The carbon cycle and CO2 over Phanerozoic time:
the role of land plants. *Phil. Trans. R. Soc. Lond. B. Biological Sciences* 353:
75-82.

April, R. & Keller, D. (1990). Mineralogy of the rhizosphere in forest soils
of the eastern United States. *Biogeochemistry* 9: 1-18.

Griffiths, R. P., Baham, J. E., & Caldwell, B. A. (1996). Soil solution
chemistry of ectomycorrhizal mats in forest soil. *Soil Biol. Biochem.* 26:
331-7.

Berner, R. A. & Cochran, M. F. (1998). Plant-induced weathering of
Hawaiian basalts. *Journ. of Sedimentary Research* 68: 723-6.

van Breeman, N., Finlay, R., Lundstrom, U., Jongmans, A. G., Giesler,
R., & Olsson, M. (2000). Mycorrhizal weathering: a true case of plant
nutrition? *Biogeochem.* 49: 53-67.

Ward, P. (2009). *The Medea Hypothesis. Is Life on Earth Ultimately Self-Destructive?* Princeton Univ. Press, Princeton, N. J., Oxford, U. K., p. 86.

Drever, J. I. & Zobrist, J. (1992). Chemical weathering of silicate rocks as a function of elevation in the southern Swiss Alps. *Geochim. Cosmochim. Acta* 56: 3209-16.

Arthur, M. A. & Fahey, T. J. (1993). Controls on soil solution chemistry in a subalpine forest in north-central Colorado. *Soil Science Soc. Amer. Journ.* 57: 1122-30.

Moulton, K. L, West, J., & Berner, R. A. (2000). Solute flux and mineral mass balance approaches to quantification of plant effects on silicate weathering. *Amer. Journ. Sci.* 300: 539-70.

Beerling, D. J. & Berner, R. A. (2005). Feedbacks and the coevolution of plants and atmospheric CO2. *PNAS USA* 102: 1302-5.

Stocker, T. F., and 9 coeditors (2014). *Climate Change 2013: The Physical Science Basis: Working Group I Contribution to the Fifth Assessment Report of the Intergovernmental Panel on Climate Change.* 1,535 pp. Cambridge University Press, Cambridge, UK.

Berner, R. A. (2013). Personal communication.

Shukla, J. & Mintz, Y. (1982). Influence of land-surface evapotranspiration on the Earth's climate. *Science* 215: 1498-501.

Parr, J. & Sullivan, L. (2005). Soil carbon sequestration in phytoliths. *Soil Biology and Biochemistry* 37: 117–24. doi: 10.1016/j.soilbio.2004.06.013

Retallack, G. J. (2001). Cenozoic expansion of grasslands and climatic cooling. *Journ. of Geology* 109: 407-26.

Berner, R. A. & Caldeira, K. (1997). The need for mass balance and feedback in the geochemical carbon cycle. *Geology* 25: 955-6.

Lieth H. & Whittaker R.H. (eds.). (1975). Primary Productivity of the Biosphere. *Ecological Studies (Analysis and Synthesis)*, vol. 14. Springer, Berlin, Heidelberg, Germany. https://doi.org/10.1007/978-3-642-80913-2_1

Woodwell, G. M. (Jan. 1978). The biota and the world carbon budget. The terrestrial biomass appears to be a net source of carbon dioxide for the atmosphere. *Science* 199 (13): 141-6.

Duchesne, L. C. & Larson, D. W. (Apr. 1989). Cellulose and the evolution of plant life. *BioScience* 39 (4): 238-41.

Sillett, S. C., et al. (15 Feb. 2020). Aboveground biomass dynamics and growth efficiency of Sequoia sempervirens forests. *Forest Ecol. and Manage-*

ment 458: 117740. https://doi.org/10.1016/j.foreco.2019.117740

Bello, C., et al. (18 Dec. 2015). Defaunation affects carbon storage in tropical forests. *Science Advances* 1: e1501105: pp. 1–10.

Danovaro, R., Corinaldesi, C., Dell'anno, A., Fabiano, M. & Corselli, C. (2005). Viruses, prokaryotes and DNA in the sediments of a deep-hypersaline anoxic basin (DHAB) of the Mediterranean Sea. *Environ. Microbiol.* 7: 586–92.

Morris, R. M., et al. (13 Dec. 2002). SAR11 clade dominates ocean surface bacterioplankton communities. *Nature* 420: 806–10. doi: 10.1038/nature01240

Carini, P., White, A. E., Campbell, E. O., & Giovannoni, S. J. (7 July 2014). Methane production by phosphate-starved SAR11 chemoheterotrophic marine bacteria. *Nature Communications* 5, Article number 4346. doi: 10.1038/ncomms5346

Hathaway, W. (June 2001). The microbe factor and its role in our climate future. *Journ. of Bacteriol.* 183 (12): 3770-83. doi: 10.1128/JB.183.12.3770-3783.2001

Pidgeon, E. (2009). Carbon sequestration by coastal marine habitats: Important missing sinks. *The Management of Natural Coastal Carbon Sinks.* IUCN 47-53.

Ricart, A. M., York, P. H., Rasheed, M. A., et al. (15 Nov. 2015). Variability of sedimentary organic carbon in patchy seagrass landscapes. *Marine Pollution Bulletin* 100 (1): 476-482. https://doi.org/10.1016/j.marpolbul.2015.09.032

Haroon, M. F., et al. (2013) Anaerobic oxidation of methane coupled to nitrate reduction in a novel archaeal lineage. *Nature* 500 (7464): 567-70. 10.1038/nature12375

Penttilä, A., Slade, E. M., Simojoki, A., Riutta, T., Minkkinen, K., & Roslin, T. (Aug. 2013). Quantifying beetle-mediated effects on gas fluxes from dung pats. *PLoS ONE* 8 (8): e71454. doi: 10.1371/journal.pone.0071454

Lloyd, J. & Farquhar, J. D. (1994). 13C atom discrimination during CO2 assimilation by the terrestrial biosphere. Oecologia 99: 201-15.

Beerling, D. (2007). *The Emerald Planet: How Plants Changed Earth's History.* Oxford Univ. Press, Oxford, U. K.

Li, K.-F., et al. (16 June 2009). Atmospheric pressure as a natural climate regulator for a terrestrial planet with a biosphere. *PNAS USA*

106 (24): 9576-9. https://doi.org/10.1073/pnas.0809436106

Charlson, R. J., Lovelock, J. E., Andreae, M. O., & Warren, S. G. (16 April 1987). Oceanic phytoplankton, atmospheric sulphur, cloud albedo and climate. *Nature* 326: 655–61. doi: 10.1038/326655a0. http://www.nature.com/nature/journal/v326/n6114/abs/326655a0.html

Andreae, M. O., et al. (1995). Biogenic sulfur emissions and aerosols over the tropical South Atlantic, 3. Atmospheric dimethylsulfide, aerosols and cloud condensation nuclei. *Journ. Geophys. Res.* 100: 11335-56. doi: 10.1029/94JD02828. http://www.agu.org/pubs/crossref/1995/94JD02828.shtml

Cropp, R.A., Gabric, A.J., McTainsh, G.H., Braddock, R.D., & Tindale, N. (2005). Coupling between ocean biota and atmospheric aerosols: Dust, dimethylsulphide, or artifact? *Global Biogeochem. Cycles* 19: GB4002. doi: 10.1029/2004GB002436. http://www.agu.org/pubs/crossref/2005/2004GB002436.shtml

Vallina, S. M., Simo, R., Gasso, S., De Boyer-Montegut, C., del Rio, E., Jurado, E., & Dachs, J. (2007). Analysis of a potential "solar radiation dose-dimethylsulfide-cloud condensation nuclei" link from globally mapped seasonal correlations. *Global Biogeochem. Cycles* 21: GB2004. doi: 10.1029/2006GB002787. http://www.agu.org/pubs/crossref/2007/2006GB002787.shtml

Shaw, G. E., Benner, R. L., Cantrell, W., & Clarke, A. D. (1998). The regulation of climate: A sulfate particle feedback loop involving deep convection--An editorial essay. *Climate Change* 39: 23-33. doi: 10.1023/A:1005341506115. http://www.ingentaconnect.com/content/klu/clim/1998/00000039/00000001/00164320

Savoca, M. S. & Nevitt, G. A. (18 March 2014). Evidence that dimethyl sulfide facilitates a tritrophic mutualism between marine primary producers and top predators. *PNAS USA* 111 (11): 4157-61. https://doi.org/10.1073/pnas.1317120111

Pfaller, J. B., et al. (9 March 2020). Odors from marine plastic debris elicit foraging behavior in sea turtles. *Current Biol.* 30 (5): PR213-R214. doi: https://doi.org/10.1016/j.cub.2020.01.071

Treguer, P., et al. (1995). The silica balance in the world ocean: A reestimate. *Science* 268 (5209): 375–9. doi: 10.1126/science.268.5209.375. PMID 17746543

Roubiex, V., et al. (June 2008). Diatom succession and silicon removal from freshwater in estuarine mixing zones: From experiment to modelling. *Estuarine Coastal and Shelf Science* 18 (1): 14-26. doi: 10.1016/j.ecss.2007.11.007

Segar, D. (2012). *Introduction to Ocean Sciences*, 3rd Ed., Ch. 5. http://www.reefimages.com/oceans/SegarOcean3Chap05. ISBN 978-0-9857859-0-1. Corpus ID: 129389136

Wigley, T. M. L. (1983). The preindustrial carbon dioxide level. *Climatic Change* 5: 315-20. https://doi.org/10.1007/BF02423528

Froschauer, A. & Coleman, J. (17 Jan. 2012). *North American bat death toll exceeds 5.5 million from white-nose syndrome.* White-Nose syndrome. org. U. S. Fish & Wildlife Service.

Gatti, L.V., Cunha, C.L., Marani, L. et al. (2023). Increased Amazon carbon emissions mainly from decline in law enforcement. *Nature* 621, 318–323. https://doi.org/10.1038/s41586-023-06390-0

O'Neill, B., Oppenheimer, M., Warren, R. et al. (2017). IPCC reasons for concern regarding climate change risks. *Nature Climate Change* 7: 28–37. https://doi.org/10.1038/nclimate3179

Doney, S. C. (March 2006). The dangers of ocean acidification. *Scientific Amer.* 3: 58-65. (secondary literature).

Fabry, V. J., et al. (April 2008). Impacts of ocean acidification on marine fauna and ecosystem processes. *ICES Journ. of Marine Science* 65 (3): 414-32. https://doi.org/10.1093/icesjms/fsn048

Zhili, H., et al. (2012). The phylogenetic composition and structure of soil microbial communities shifts in response to elevated carbon dioxide. *The ISME Journ.* 6: 259–72. doi: 10.1038/ismej.2011.99

Brooks, G. L. & Whittaker, J. B. (24 Dec. 2001). Responses of three generations of a xylem feeding insect, Neophilaenus lineatus (Homoptera), to elevated CO2. *Global Change Biol.* 5 (4): 395-401. https://doi.org/10.1046/j.1365-2486.1999.00239.x

Awmack, C. (30 Oct. 2003). Host plant effects on the performance of the aphid Aulacorthum solani (Kalt.) (Homoptera: Aphididae) at ambient and elevated CO2. *Global Change Biol.* 3 (6): 545-9. https://doi.org/10.1046/j.1365-2486.1997.t01-1-00087.x

CHAPTER 4. ORGANISMS CREATED HIGH OXYGEN LEVELS, WHICH ALLOWED COMPLEX LIFE TO EVOLVE AND DIVERSIFY

Life Increased Earth's Oxygen Level in the Great Oxidation Event

We saw in Chapter 2 that life profoundly changed the chemistry of the atmosphere in ways favorable to life and biodiversity. One of the most important methods by which it did this was the creation and maintenance of an atmosphere high in oxygen. Oxygen concentrations in the Earth's atmosphere rose from negligible levels of much less than 1% in the Archaean Eon (4 to 2.5 bya) to about 21% in the present day, essentially all as a result of the actions of organisms. This also oxygenated the soil and both fresh and salt water ecosystems. Life, through both photosynthesis and the removal of hydrogen and carbon, produced the high level of oxygen present in today's atmosphere, and needed for higher life. Prior to the origin of life, atmospheric oxygen was negligible near Earth's surface (Kasting, 1993). For almost the entire first half of Earth's 4.6-billion-year history, the air had less than one part per million of oxygen. It was probably less than 10-5 of the level that is present today (Sessions et al., 2009). Then came the Great Oxidation Event, or GOE (ibid., and references therein), about 2.3 to 2.45 bya (ibid.), when oxygen levels increased a great amount, although only to a few percent of today's levels. Cyanobacteria evolved oxygenic photosynthesis at about the time of the GOE or a little before it. They are likely the group that produced most of the oxygen that accumulated during the GOE.

Oxygenic photosynthesis uses light from the sun to manufacture carbohydrate, which plants use as a food and energy source. In so doing, oxygenic photosynthesis produces oxygen. Oxygenic photosynthesis is carried out primarily by plants, algae, phytoplankton, and cyanobacteria. It has been the main source of oxygen in the atmosphere since it evolved. Today it produces 98 to 99% of the atmosphere's oxygen. There is also photosynthesis that does not produce oxygen, called anoxygenic photosynthesis. Henceforth in this chapter, when I say photosynthesis, I will mean oxygenic photosynthesis. Oxygen production by photosynthesis is necessary to build an atmosphere with high levels of oxygen, but is not sufficient to do so. Respiration by organisms, the decay of dead organisms, and fire, consume oxygen, with the result that there is no net gain of oxygen from photosynthesis alone. Carbon combines with oxygen and so depletes it during these processes. And carbon and hydrogen react with oxygen and consume it in other reactions. Burial of carbon dioxide regulates temperature, but does not increase atmospheric oxygen levels, since one oxygen molecule, which consists of two oxygen atoms, is buried with each carbon atom when carbon dioxide (CO_2) is buried. To increase free atmospheric oxygen, something must remove hydrogen or carbon atoms that are not attached to oxygen, in addition to producing oxygen by photosynthesis. This is most commonly accomplished by removing carbon, rather than hydrogen. The chemistry is a little more complex, but for the purposes of this discussion, what I will call reduced carbon is carbon that is not combined with oxygen. Reduced carbon is usually combined with hydrogen, so burying it generally helps increase oxygen in the air in two ways: burial of carbon and burial of hydrogen. When carbon or hydrogen that are not combined with oxygen are buried, there is less of them to combine with and deplete oxygen, so there are more free oxygen molecules that are not combined with either of these elements in the biosphere. This way, the concentration of oxygen in the atmosphere increases.

Oxygen was toxic to many prokaryotes when photosynthesis first appeared. It could kill them. There are prokaryotes today that can only live in habitats where there is no oxygen. So a sudden rise in oxygen certainly decreased in the diversity of prokaryotes. It could have potentially been catastrophic to diversity at that time, and some think the GOE caused a mass extinction of prokaryotes, providing a counter example to the ABH/Pachamama Hypothesis. Prokaryotes and viruses were the only life at the time of the GOE.

But Cardona (2018) presented evidence that oxygenic photosynthesis evolved 3.6 bya, a billion years earlier than previously thought. There was a time lag between the first appearance of oxygenic photosynthesis and the GOE. Oxygen levels built up and then went down several times because compounds of sulfur, iron, manganese, nitrogen (mainly as ammonia), and other elements combined with the oxygen every time its levels built up, keeping its concentration in the air low for a long time (Kump et al., and references therein). Oxygen could accumulate only after it fully oxidized these compounds, making it so that they could no longer combine with oxygen. This is why there was a long time between the beginning of photosynthesis and the GOE. Weathering of rocks also removed a great amount of oxygen from the air in the past, as it does today. The transformation to an atmosphere high in oxygen was a gradual process that took several hundreds of millions of years, and there were continual small rises in oxygen levels followed by small decreases as oxygen combined with the reduced chemicals mentioned above and rocks were weathered. Prokaryotes have very short generation times and increase their populations very quickly, so they can adapt and evolve very rapidly. So prokaryotes had the ability and ample time to adapt to the increase in oxygen, and were adapted to it when the GOE came. They evolved enzymes that protect cells from toxic oxygen, preventing it from damaging them; these enzymes still exist today. There also would have been areas below the seafloor where prokaryotes could live where there was no oxygen, as there are today. Some microbe species may have gone extinct, and some may have decreased in abundance for a while, but it is highly unlikely that the increase in oxygen caused a mass extinction of microbes.

In fact, even the short-term effects of the GOE promoted diversity. Some prokaryotes suddenly had a way to take advantage of the many niches created by sulfides, ammonia, iron in its form that readily combines with oxygen, and other compounds that easily combine with oxygen that were abundant as a result of photosynthesis. These prokaryotes diversified, with some species evolving that oxidized sulfides, others oxidizing ammonia, others oxidizing iron, and others evolving that oxidized yet other compounds. These bacteria created several new compounds, such as nitrate, sulfate, and iron and manganese oxides, that comprised new niches that several species of microbes evolved to exploit, and these latter forms also diversified. Still other new niches opened up with the sudden abundance of readily available reduced carbon compounds produced by oxygenic photosynthesis that were available at a low energy cost, leading to the evolution and diversification of species of prokaryotes that could utilize these niches by consuming these compounds. Then other prokaryotes evolved that could eat these microbes, and they diversified. Many species of virus evolved that became parasites and symbionts of all of these new prokaryote species. Said another way, viruses diversified and used the many new niches created by these many new prokaryote species. Thus, oxygenic photosynthesis directly and indirectly created several new niches for many species of prokaryotes and viruses, allowing both to diversify tremendously.

The high oxygen content caused a key innovation: the evolution of cellular respiration. The chemicals that evolved to protect the prokaryotes from oxygen evolved further into the chemicals that carry out cellular respiration, the "breathing" of cells. This is a major evolutionary innovation that allows cells, including microbes, to utilize oxygen as an energy source. It is what makes all complex multicellular life, including trees, elephants, dinosaurs, whales, and humans, possible. The high oxygen level in the air and life's ability to use it for energy, both created by life, enabled the evolution, survival, and diversification into myriad species and higher taxa of complex multicellular. This is because oxygen provided the necessary energy for complex multicellular life to evolve, and cellular respiration provided the mechanism to utilize it.

Sulfate-reducing bacteria, which are bacteria that get their energy from sulfur, had a major role in the GOE (Halevy et al., 2012). They produced sulfide, which combined with iron, making the iron unable to combine with oxygen, and thus reduce the amount of it. They also produced oxygen.

Remember, to increase atmospheric oxygen, carbon or hydrogen not combined with oxygen must be removed from the biosphere. It appears it was the burial of hydrogen, along with photosynthesis, had a major role in increasing oxygen during the GOE. Hydrogen cannot rise into the air because it easily combines with other chemicals. Catling et al. (2001) proposed that prokaryotes called methanogens converted hydrogen into methane, which contains four atoms of hydrogen and one of carbon. It is light in weight and does not readily combine with others chemicals, so rose it up to the stratosphere, where UV light from the sun broke it into carbon and hydrogen. Since hydrogen is light, much of it escaped into space, while much of the heavier carbon stayed in Earth's atmosphere. Pope et al. (2012) provided evidence that this occurred during the GOE.

Hoehler et al. (2001) showed that today's microbial mats that include cyanobacteria on the coast of Mexico's Baja Peninsula pump a thousand times more hydrogen into the air than typical volcanoes and hydrothermal vents, the two nonbiological hydrogen sources. Microbial mats at the time of the GOE would have behaved similarly. Catling's group thinks methane built up in the atmosphere to concentrations a few hundred to a few thousand times greater than modern levels; therefore, copious quantities of hydrogen would have escaped into space by this mechanism. It is thus likely that microbial mats and methanogens were primarily responsible for a good deal of the early oxygen build up in Earth's atmosphere via hydrogen removal.

High Oxygen Levels Diversified Minerals, Increasing Biodiversity

The GOE also caused a tremendous increase in the variety of minerals on Earth. It is estimated that the GOE was directly responsible for about 2,900 new minerals on Earth, by combining already existing minerals with oxygen (Hazen et al., 2008). After the GOE, chemical elements, such as uranium, vanadium, manganese, copper, selenium, tellurium, arsenic, antimony, bismuth, silver, and mercury, could be present in one or more forms combined with oxygen in minerals near Earth's surface. Also, the relative abundances of clay minerals were enhanced by major biological events, including the rise of deep-rooted vascular plants and their symbiotic fungi.

It is not only photosynthesis that increased the number of minerals. Bacteria interact with metal ions more than any other type of life does. They have played and still play a large role in the minerals of most soils and sediments. Many of life's most important essential elements, such as carbon, sulfur, nitrogen, and phosphorous, are abundant, but to a large degree unavailable to many prokaryotes and eukaryotes. Some bacteria make these elements available to many other life forms, and use them to make a great range of minerals. It has been suggested that they clean natural environments of metals that are toxic when in concentrations that are too high, such as cadmium, nickel, and copper. Microbial communities work together to accomplish a great deal of these functions (Douglas and Beveridge, 1998).

The life-induced increase in mineral diversity, like an increase in biodiversity, is an increase in information content on Earth, caused by life. We know that prokaryotes often partition mineral niches, with different species specializing on different mineral compounds. It is thus reasonable to assume that the increase in minerals led to an increase in prokaryote diversity, as prokaryotes diversified and filled the new mineral niches created by life, in a life-induced positive feedback loop between life and minerals. This is another example of the coevolution of the biosphere and geosphere.

High Oxygen Levels Increased the Number of Amino Acids Life's Proteins Use, Increasing Biodiversity

Amino acids are the building-block molecules used to make proteins in all life. High atmospheric oxygen levels were the main reason that the number of amino acids used to make proteins by life was increased from 13 to 20, even though functional proteins can be assembled by the original 13 amino acids alone (Granold et al., 2018). Oxygen caused the formation of free radicals, which the seven new amino acids, which had greater reactivity than the original 13, reacted with. Increasing the number of amino acids that organisms used to make proteins greatly increased the variety of proteins life could and did make. This profoundly increased evolutionary innovation and diversification into new life forms, amplifying biodiversity.

Life Caused Further Rises in Oxygen Levels after the GOE

In the time since the GOE, reduced carbon was buried by phytoplankton, trees, echinoderms, salps, and other organisms, by the mechanisms described in Chapter 3. This happened while plants, phytoplankton, cyanobacteria, and other photosynthesizers added oxygen to the air by photosynthesis. Thus, life was almost the sole cause in the rise of oxygen levels after the GOE. Of course, organisms sequestered oxidized carbon too, but this did not increase atmospheric oxygen levels. Nonbiological forces also removed some reduced carbon and hydrogen from the system, but much less than life did. And nonbiological factors have removed a great deal of oxygen from the atmosphere in the past because substances like iron, sulfur, manganese, and so on combined with it.

The oxygen content of the early Earth's surface environment is thought to have permanently increased in two broad steps: the previously discussed GOE, and the Neoproterozoic Oxidation Event (NOE), during which oxygen possibly accumulated to the levels required to support animal life and ventilate the deep oceans. The Neoproterozoic Era was from one bya to 541 mya. While the GOE is well accepted, the timing and extent of the NOE is not well agreed upon. Och and Shields-Zhou (2012) reviewed the evidence for the NOE and presented convincing evidence that it occurred and was important, and was instrumental in major biological innovations and an abrupt diversification of the biota (organisms) during the Ediacaran Period of about 635–541 mya. The Ediacaran biota were the first multicellular organisms, although many of the animals could not move, and were attached to rocks or the seafloor. They included trilobites and jellyfish, which are more properly called sea jellies, since they are not fish. They were not as complex on the average as the animals that followed them. Canfield et al. (2014) also presented convincing evidence that oxygen levels rose about 580 mya to 15% of today's levels, or about 3% of the atmosphere. This corresponds well in time with the proposed event called the Avalon explosion, when the Ediacaran biota are thought to have undergone a tremendous diversification into many new species 575 mya Shen et al. (2008). The timing and extent of the NOE is still a subject of debate, but these researchers made a convincing case for a rise in oxygen on a global scale during the Late Neoproterozoic Era.

Therefore, *this oxygen increase is significant because it made possible the rise and diversification of the first complex multicellular animals.* It is not certain what caused this rise in oxygen, but there are some good hypotheses. Below are what I consider to be the two best ones. In both hypotheses, life played a major role in increasing oxygen levels.

Lichens and bryophytes (mosses and their relatives) likely aided this rise

106

in oxygen by increasing weathering and thus burial of reduced carbon. Lichens and bryophytes have the ability to cause weathering Porada et al. (2014), and these groups were abundant at this time. Higher land plants had not yet evolved.

Sponges first appear in the fossil record about 650 mya and could have survived very well in atmospheric oxygen levels as low as 0.5% to 4% of present ones (Mills et al., 2014; Mills and Canfield, 2014), which probably existed well before higher animals evolved. Some modern sponges can live under oxygen levels 200 times less than present in the air today, which is about 0.105% of present levels. Sponges filter hundreds of liters of water per day, removing carbon, some of which is reduced and gets sequestered when the sponge (after its death) or its waste products are buried. A great number of sponges doing this would have cooled the planet and increased oxygen substantially. Sponges would have been abundant enough to increase oxygen levels in the sea by this mechanism substantially by the time of the appearance of the Ediacaran fauna.

Bear in mind that the increase in atmospheric oxygen preceding the rise of the Ediacaran animals could have resulted from a combination of two or more different mechanisms.

The Cambrian explosion was a huge diversification of animals that started about 541 mya, and brought about the origin of modern animals and practically all of the major animal phyla. It is when higher animals appeared and developed shells, boney skeletons, complex nervous systems, and sensory organs. There are several hypotheses as to why life suddenly became more complex and diverse at this time. Not all of the hypotheses emphasize a rise in oxygen, although oxygen had to rise sufficiently to support this great increase in complexity and diversity.

In my view, the most convincing mechanism for the increase in oxygen that precipitated the Cambrian explosion is from Logan et al. (1995). They say that bacteria thrived in the Proterozoic Eon (2,500 to 538.8 mya), consuming oxygen near the sea surface, and inhibiting the transport of oxygen to the deep ocean. Zooplankton, which are one-celled organisms that live in the shallow waters of the sea, out-competed the bacteria and greatly reduced their numbers, reducing the consumption and depletion of oxygen near the ocean's surface. There was an increase and diversification of zooplankton that formed large fecal pellets loaded with reduced carbon that sank and were buried in the sea floor. This would have buried a great deal of reduced carbon, causing an increase in oxygen in the ocean, and hence providing the high oxygen levels needed for the Cambrian explosion. The timing is just right, for zooplankton did this just prior to the Cambrian explosion.

Lenton and Watson (2004) proposed that colonization of land by algae, fungi, and other small organisms in the Neoproterozoic Era resulted in selective weathering of phosphorous from rocks, some of which was washed into the oceans. Increased phosphorous increases phytoplankton numbers, so they would have produced more oxygen through photosynthesis, and more of them would have sunk to the seafloor upon death, burying carbon, some of it reduced. This created the high oxygen levels in the sea necessary for the Cambrian explosion.

Heckman et al. (2021) provided evidence that land plants and fungi were present early enough to have produced the oxygen needed for the Cambrian explosion by weathering and burying reduced carbon.

Cyanobacteria remained the principal oxygen producers throughout almost all of the Proterozoic Eon. Green algae, which are eukaryotes, joined cyanobacteria as major photosynthesizers and producers of oxygen on the continental shelves of the seas near the end of this eon. The diversification of the four major groups of phytoplankton--dinoflagellates, coccolithophorids, foraminiferans, and diatoms—occurred in the Mesozoic Era (252 to 66 mya). These became important oxygen generators from the time of their evolution until the present. They are the major phytoplankton that produce most of the oxygen and support the major food webs in today's oceans. Cyanobacteria are still important oxygen generators in the ocean. Of course, land plants, especially trees, have been important producers of oxygen ever since they first appeared. Today, cyanobacteria, phytoplankton, and land plants produce almost all of Earth's oxygen. Land and sea plants produce 105 billion tons of biomass per year (Beerling, 2007), indicating they have a profound effect on regulating carbon dioxide, reduced carbon, and climate, and produce a great deal of oxygen. Scientists are debating the relative importance of each group of photosynthetic species, and what follows are approximations. Freshwater and marine phytoplankton are responsible for over half of the Earth's total productivity and oxygen production, although they are less than 1% of the Earth's photosynthetic biomass (ibid.). Land plants account for 20 to 45% of Earth's oxygen, and constitute 90% of the photosynthetic biomass (ibid.). Cyanobacteria account for most of the remainder of the Earth's oxygen production.

The increase in atmospheric oxygen, eventually to 21%, created by life, provided the energy source necessary for all higher life, which could not exist without it. Without high atmospheric oxygen levels, all life would be viruses or simple organisms with one or a few cells. High oxygen in both the air and water is necessary for the evolution and existence of all vertebrates, invertebrates, higher plants, and fungi.

Lenton (2012) pointed out that the origin and spread of primitive plants such as algae and moss progressively increased weathering of phosphorus from rocks. This released phosphorus into the sea, leading to a rise in atmospheric oxygen from about 12% by volume 570 mya, plus or minus 40 mya, to its present level of about 21% by volume by about 420 to 400 mya, by the mechanisms discussed earlier. His experiments reveal that spreading-leaved Earth moss (*Physcomitrella patens*) amplifies phosphorus weathering by a factor of up to 60! Lenton et al. (2016) up-dated this to assert that land plants increased oxygen to today's levels by as early as 420 to 400 mya, and with this came the regulation and stabilization of atmospheric oxygen levels by fires, which continued to this day.

Weathering by higher land plants and their symbiotic fungi and the burial of these plants about 400 mya caused an increase in oxygen (Dahl et al., 2012) that facilitated the rise and diversification of large predatory fish, which required high oxygen. Some were up to about 9 meters (about 30 feet) long. This shows that biologically-induced increases in oxygen caused major breakthroughs in animal evolution. It provides another example of the coevolution of life and the physical/chemical environment. *If the portion of oxygen that was supplied to the atmosphere by the gradual burial of plant organic matter over the eons were removed, the remaining oxygen content would be an asphyxiating 10% today, equivalent to the thin atmosphere at about 5.5 km (about 3.4 miles) above sea level!*

Atmospheric oxygen skyrocketed in the Carboniferous Period (about 358.9 to 298.9 mya) because of the great number of very large plants buried in swamps, making the coal beds that fueled the Industrial Revolution. The 35% oxygen content of the air at oxygen's peak levels during this period greatly increased the frequency and intensity of fires because fire requires oxygen to burn. The higher the oxygen level in the air, the greater the number and size of fires. Lignin evolved as a defense against fire. Also, thick, bark-like, corky layers on the outside of plants of the Carboniferous Period evolved for fire resistance. These thick layers allowed trees and tree-like plants to become very tall. Plants that evolved this defense included trees distantly related to modern club mosses, which are small today, but grew to over 40 meters (about 131 feet) tall at this time. Tall plants also included species of tree-like horsetails reaching over 30 meters (about 98.5 feet) in height, and they are only 1.5 meters (about 5 feet) tall today. These tall forms provided new habitats for insects and other animals to use for food and shelter. There was a sequence of change and diversification of life driven by life, as follows: plants were buried in swamps in great numbers; this buried tremendous amounts of reduced carbon; oxygen gradually increased to levels that peaked at 35%; fires gradually increased to great sizes and frequencies; lignin and thick, bark-like, corky layers evolved for fire resistance; this gave support that allowed some plants to get extremely tall, increasing plant diversity; and animals diversified as they took advantage of the towering three-dimensional habitat the tall plants provided. Plants thus drove their evolution and diversification and that of several animal species.

There was a sequential chain and positive feedback loop that involved actions by plants, evolution, and increased oxygen, as follows (Berner, Berner, and Moulton, 2023). Weathering by and burial of plants caused more oxygen, more atmospheric oxygen causes more fires, which selects for fire resistance, including the ability to be charred and not burn. This produces charcoal, which tends to be buried as reduced carbon, increasing atmospheric oxygen, and hence more fires. The existence of more fires selects for more fire resistance, producing more charcoal, and so on. This stopped when a stable state was reached at about 35% oxygen in the air.

The 35% level of oxygen in the atmosphere allowed arthropods to become gigantic. The respiratory system of insects and many other arthropods functions by passive diffusion. They have no active pumps, such as lungs, to breathe with. So the oxygen content of the atmosphere limits their size. The high oxygen level and tall plants allowed arthropods to evolve into gigantic and spectacular forms. Huge insects evolved. The largest insect ever known to live, an ancestor of the dragonfly, with a wingspan of 73 centimeters (almost 2 feet, 5 inches), appeared at this time. There were centipedes and millipedes over 1 meter (3 feet, 3 inches) long, including a millipede over 2 meters long (about 6 feet, 7 inches) and a half meter (about 1 foot, 8 inches) wide. There were huge spiders and scorpions. The diversity of arthropods probably increased quite a bit because small and medium-sized arthropods would have continued to exist as well. Although amphibians breathe actively with lungs, they also frequently breathe passively through the skin, so oxygen can limit their size. Thus, the high oxygen levels allowed them to become very large at this time, and some did, reaching 5 meters (about 16 feet, 5 inches), with amphibian diversity likely increasing for the same reason as arthropod diversity did. By the end of the Permian Period, which ended about 251.9 mya, oxygen levels were down to 15%, and arthropods and amphibians were no longer so large. Extreme fluctuations in oxygen need not lead to catastrophe, and extreme levels returned to intermediate ones in time. The increase in oxygen in the Carboniferous Period caused by life is an example of biology apparently causing an increase in the diversity of insects, amphibians, and some plants, supporting the ABH/Pachamama Hypothesis.

Falkowski et al. (2005) modeled oxygen concentrations, finding that oxygen levels rose gradually and with fluctuations from about 10% to today's 21% level, from 205 mya onward, with pronounced rises in the early Jurassic Period (about 200 mya) and Eocene Epoch (about 55 mya). They suggested that a primary reason for the overall increase in oxygen was the drifting apart of the continents around the Atlantic Ocean, sequestering reduced carbon by nonbiological weathering. This was necessary for and facilitated the evolution, increase in average size, diversification, and rise to prominence of large mammals, which require high atmospheric oxygen. These events happened to mammals at about the time of the increase in oxygen levels during the Eocene Epoch and immediately after it. This is also when placental mammals became prominent. Placental mammals all bear live young, and carry the fetus in the uterus of the mother to birth, setting them apart from marsupials that keep their young in pouches and primitive mammals like the duck-billed platypus. They contain the vast majority of living mammals, including wolves, whales, lions, monkeys, and humans. Organisms also supplied oxygen through photosynthesis and buried reduced carbon. This co-evolution of geology and biology only partially supports the Pachamama Hypothesis, since the authors think the increase in oxygen was mainly done by a nonbiological process.

The Amazon River and Phytoplankton
Together Produce Great Quantities of Oxygen

An interaction of the Amazon rainforest and ocean phytoplankton produce a great portion of Earth's oxygen. The Amazon Basin has at least one-tenth of the Earth's species. It has trillions of leaves, and each tree releases hundreds of thousands of cubic meters of oxygen in its life. It produces a fifth of Earth's oxygen. But it is almost a closed system, so almost all the oxygen gets reabsorbed in the forest through respiration and decomposition. Half of the Amazon's rain is created by its trees. The Amazon River system supplies 20% of global water runoff to the seas. This water carries soil with it, totaling two million tons of sediment per day. Vast amounts of leaves, branches, and nutrients are carried in this sediment. It travels through the Amazon delta into the Atlantic Ocean, where the great quantity organic material and nutrients in the river provide nutrition for phytoplankton to bloom near the sea's surface. The bloom spreads out to sea, doubling in one day, growing to cover a massive 25,000 square miles of sea surface, carrying out photosynthesis, absorbing carbon dioxide, and producing billions of gallons of oxygen each day. In as little as three days, almost all of the plankton that are not eaten die, some sinking to the seafloor, where they get buried, sequestering carbon, some of it reduced. The phytoplankton stay at the seafloor for millennia, making a blanket a half mile thick in some places. This provides the Earth with a huge net gain in oxygen. Wherever other rainforests have rivers that run to the sea, there are similar interactions of forests, rain, rivers, and phytoplankton that produce large amounts of oxygen. This a major mechanism by which phytoplankton produce over half the planet's oxygen. Of course, rainforests indirectly make large amounts of oxygen as well by this mechanism. It is the interaction of forests, rainfall, rivers, sea, and phytoplankton that produce large quantities of oxygen, and the rainforest and phytoplankton together, not rainforests alone, can be called the lungs of the Earth.

Life's Production of High Oxygen Levels Helps Diversity in Many Ways Besides Providing an Energy Source for the Evolution and Diversification of Complex Life

The production of high oxygen levels by life led to the formation of the ozone layer. Ozone (O3) is three atoms of oxygen bonded together into one molecule, and is formed from molecular oxygen (O2), which is two oxygen atoms bonded together. Oxygen normally occurs as O2. The ozone layer formed from the high atmospheric oxygen levels created by life. So life is the ultimate cause of the ozone layer. It is in the lower portion of the stratosphere, about 13 to 20 kilometers (about 8.1 to 12.4 miles) above Earth. This layer protects the Earth from high levels of ultraviolet (UV) radiation from the sun, absorbing 97-99% of the sun's damaging medium-frequency UV radiation. Without the ozone layer, UV radiation would prevent the evolution and existence of any life except the simplest forms, and those underground or deep enough in the sea to be shielded by water from UV. UV causes genetic damage by inducing mutations in the DNA, weakens the immune system, harms the eyes and can cause blindness, destroys vitamin A in the skin, can accelerate aging of the skin by damaging collagen fibers, is the main cause of sunburn, creates carcinogenic and cell-damaging free radicals, and can induce skin cancer, including the deadliest form of it: malignant melanoma. These mechanisms of harm are not limited to humans, but affect animals, plants, fungi, and microbes. Therefore, the ozone layer protects essentially all life above the ground and in the upper part of the sea. The formation of the ozone layer allowed phytoplankton that previously lived deeper in the ocean to move up to the surface, and hence better photosynthesize and increase their populations. They could thus support many more animals in the ocean's food web. They could also produce more oxygen and sequester more carbon. So, they increased atmospheric oxygen and hence enhanced the ozone layer, in a positive feedback loop that eventually reached an equilibrium and stabilized. The ozone layer also allowed plants to colonize land. Land plants photosynthesized and buried reduced carbon, increasing oxygen and ozone, in another positive feedback loop.

Sulfate-reducing bacteria produce hydrogen sulfide, which is toxic to almost all life. These bacteria cannot live in the presence of oxygen. Life's production of a high-oxygen atmosphere confines these bacteria to small, limited habitats where there is no oxygen, greatly limiting toxic hydrogen sulfide production. There are also microbes that obtain their food by oxidizing, and thus destroying, hydrogen sulfide; this is another way life limits the amount of hydrogen sulfide. Oxygen also limits methanogens, microbes that produce the potent greenhouse gas, methane, because these microbes cannot live in the presence of oxygen. Oxygen also combines with and hence destroys methane. The sulfide-reducing bacteria and methanogens add to life's diversity, while being prevented by life from reducing it

Oxygen and Fire are Regulated by
Negative Feedback Mediated by Life

High oxygen, working in concert with lightning, also benefits life by the creation of fire. Natural fires need oxygen, lightning, and fuel, such as trees, to burn. Lightning is constantly starting forest fires randomly around the world. Before human effects became large, fire burned over 19 million square miles on Earth each year. Within hours, a spark from one lightning bolt turns acres of forest to flames. This renews life. Forests are full of dead and diseased trees that lock up life's nutrients, such as carbon, nitrogen, sulfur, and phosphorous. Fire releases and recycles these nutrients and fertilizes the soil. Fire can shorten decomposition from decades in some forests to hours, making nutrients available immediately. It allows the biosphere to turn over more efficiently and effectively. A burst of new growth follows a fire almost immediately. Large burned areas are rejuvenated within months. Fire thus converts dead organisms to living ones.

In many areas, only forests would exist without fire. Fire creates a mosaic of many different ecosystems because it burns some of the forest to earlier stages of succession. In ecological succession, more complex stages replace earlier stages. For example, lichens are replaced by grassy meadows, which are replaced by shrubs, which are replaced by various species of trees, which are replaced by other species of trees of the final stage. Fires can burn forests back to each of these stages of ecological succession. Thus, multiple fires of different sizes and intensities can result in many different stages of ecological succession existing simultaneously. In the greater Yellowstone ecosystem, for example, periodic forest fires ensure there are both forests and meadows, instead of only forests. Although forests are in general more diverse than meadows, the overall diversity is increased when both types of ecosystem are present, rather than just one. Over large areas, multiple fires can greatly increase overall diversity by often causing many or all stages of succession to exist simultaneously.

Roche et al. (2019) found forest fires conserve water. Natural wildfires in the 5,310-square-kilometer (about 2,050-square-mile) American River basin in California save the forest about 773 billion liters of water per year, because trees emit water. With fewer trees losing water, trees can grow larger, the forest is wetter and healthier, and can support a greater diversity of animals.

Fires are essential for the health of many terrestrial ecosystems. They control understory growth, preventing hot, destructive fires that would occur if the understory became too dense. Fire sometimes increases diversity by preventing the plant community from reaching a state where plants are crowding each other and competing for space, sunlight, and nutrients. Preventing them from reaching this state prevents the best competitors from eliminating the weaker competitors from the plant community. This results in more plant species living in the community. And that allows more animal species to exist there as well.

Fires in Africa supply up to half of the phosphorous annually to the Amazon basin (Barkley et al., 2019), as winds blow the phosphorous to the Amazon, helping its productivity and diversity. African fires are also an important source of soluble phosphorous to the Tropical Atlantic Ocean and oceans in the Southern Hemisphere and may be important for marine productivity, particularly in the northern summer and fall (ibid.). The increased phosphorous leads to phytoplankton blooms, feeding the countless animals in the marine food webs.

Wildfires that burned 21% of Australia's temperate and broadleaf forests in 2019 caused iron and phosphorous to be delivered to the southern Pacific Ocean thousands of miles away (Tang et al., 2021). The nutrients spurred enormous phytoplankton blooms that grew over 2,000 miles wide, an area larger than Australia, concentrated in patches south of Australia and far off the western coast of South America. This supplied food to many animal species in the food web, and reduced greenhouse gasses by enabling the burial of carbon, keeping the climate cooler and more favorable to life in general.

The Thomas Fire that burned from December 5, 2017 until January 12, 2018 in Ventura and Santa Barbara Counties, California, fertilized the Santa Barbara Channel, resulting in an abnormally high number of dinoflagellates (Kramer et al., 2020), supporting the food web there. Wildfires also fertilize and cause algal blooms in freshwater rivers and lakes.

However, too much fire can decrease diversity and be destructive to sea life. In 1997, wildfires in Indonesia fertilized a coral reef in Sumatra with too much nutrient, causing a red tide that killed the reef and its fish. Such large fires do not normally occur in the absence of human effects, such as global warming.

Life Stabilized Atmospheric Oxygen Levels for the Last 350 Million Years with Imperfect Negative Feedback

Atmospheric oxygen has been remarkably stable over the past 350 million years (Lenton and Watson, 2000). The range of atmospheric oxygen has been kept between 17 and 35% during this time. This is strong evidence for effective regulation of atmospheric oxygen at close to optimal levels for life by some mechanism. Too little oxygen would make complex, multi-cellular life impossible, while too much is toxic to cells. Too little oxygen would allow too few fires, and too much of it would allow too many fires and fires that are too large, to maximally aid life. The 21% oxygen level of today's atmosphere is as near optimal for life as one could hope for, for both of these reasons. The regulation of atmospheric oxygen occurs on million-year time scales and is accomplished in large part by regulation of the burial of reduced carbon and weathering rates (ibid.), and life is involved in this.

If a negative feedback mechanism driven by life regulated atmospheric oxygen, keeping it at levels favorable to life, this would strongly support both the Gaia Hypothesis and the Pachamama Hypothesis. Such stabilizing negative feedback exists. One negative feedback loop works as follows. Increased oxygen causes more fire, which reduces the oxygen level. Fire consumes and hence depletes oxygen. Fire also reduces the amount of forest, which means less weathering by tree roots and their symbiotic fungi, and hence less reduced carbon burial. This means that there is more reduced carbon in the air. The increased reduced carbon combines with and thus further decreases the amount of atmospheric oxygen. So there is a negative feedback loop that stabilizes atmospheric oxygen and involves life. Another more complete model involving a negative feedback loop connects fire with land and sea ecosystems and fire (ibid.). Phosphorus is usually the ultimate limiting nutrient in the global ocean. It is supplied to the surface ocean partly by the upwelling of deeper water, but on geological time scales, the source of bio-available phosphorus to the sea is the weathering of continental rocks, greatly accelerated primarily by land plants and their fungi. Less oxygen means less fire, and hence more plants, including trees. More plants mean more weathering of continental rocks, so more phosphorus is released from the soil. This is delivered to the sea via rivers. Phosphorus causes increased growth of phytoplankton, which produce more oxygen by their photosynthesis. Also, when there are more phytoplankton, more of them die and sink to the bottom of the sea. More carbon, some of it reduced, is buried in the seafloor as a result, so there is more oxygen in the biosphere. On the other hand, an increase in oxygen means more fire and hence less vegetation, so less phosphorus is weathered, resulting in less oxygen. This is a negative feedback loop that stabilizes oxygen levels in the air and ocean, keeping them at levels favorable to life, over long time periods. It is accomplished by life, supporting the Pachamama Hypothesis.

However, the negative feedback is not foolproof and can be over-ridden by catastrophe, such as a large asteroid strike or massive volcanism. Nor does it keep atmospheric oxygen levels perfectly stable at all times. They reached 35% in the middle of the Carboniferous, and this was caused by life. However, the biologically-mediated negative feedback has always been restored, and, with the help of life, oxygen levels have remained relatively stable and favorable to life for 350 million years.

Notably, this system also features a life-induced negative feedback loop that keeps fire at intermediate levels favorable to life. More fire reduces oxygen by consuming it, and reduces vegetation by burning it. Fire needs oxygen and fuel in the form of vegetation. With less of these, there is less fire. Conversely, less fire means more oxygen and fuel, so more fire. All of the negative feedback loops discussed here show that fire and oxygen control each other in a stabilizing system utilizing biology—the trees and phytoplankton.

Some animals also help keep fire close to optimal levels. Burrowing animals such as small marsupials reintroduced to forests in Australia from which they had been driven locally extinct reduced leaf litter by 24% by burrowing and turning over the dead leaves with soil, helping break the litter down (Hayward et al., 2016). The reduction in leaf litter reduced the intensity and rate of spread of fires, so increased the number of species in the ecosystem.

Experiments and the fossil record show that methanogens also control oxygen, because the methane they produce combines with oxygen, removing it from the air (Watson et al., 1978). Today, methanogens play a key role in preventing oxygen from getting to levels that are higher than optimal for life. Each year, about 1014 moles of methane are produced, primarily by methanogens (ibid.). Methane removes oxygen in the lower atmosphere at about 2,000 megatons per year (ibid.). Without methanogens burying a great quantity of reduced carbon, atmospheric oxygen concentration would rise by 1% in as little as 12,000 years, a geologically short time. The authors of this study suggest that oxygen has been regulated by methane, and perhaps by nitrous oxide and other chemicals, keeping it at about 10–25% of the atmosphere for very long periods. Methanogens may regulate oxygen by negative feedback. The higher the oxygen level, the more methane combines with and depletes it. This feedback is not perfect. This regulation of atmospheric oxygen, keeping it at levels favorable to life, is done by life (methanogens).

Likewise, photosynthesis produces oxygen, which removes the potent greenhouse gas methane when the oxygen combines with it. Oxygen also controls methane by controlling the growth of the methanogens that produce methane, because they cannot grow in the presence of oxygen. Thus, methanogens and photosynthetic organisms help biodiversity by regulating both oxygen and methane, countering each other, and keeping these gasses at levels favorable to life. Thus, as hypothesized by the Pachamama Hypothesis, life helps life and fosters biodiversity.

The GOE May Have Caused a Glaciation, but Almost Surely not a Mass Extinction

The rise of oxygen profoundly increased biodiversity and the complexity of life in the long run, but the GOE may have decreased biodiversity for as much as 300 million years, although it did not cause a mass extinction. Oxygen combined with methane, depleting it to the point that it reached very low levels. Methane had been helping keep Earth warm, and its loss made the planet so cold that ice formed over large areas. The regulation of methane and oxygen by each other discussed above is not always fully effective. The large amount of ice coverage raised the reflectivity of the Earth, causing heat to be reflected from the planet into space, which caused further cooling and ice formation, in a positive feedback loop. This made Earth extremely cold. Some researchers think that a great deal of the Earth froze, triggering the Huronian glaciation, which occurred right after the GOE, lasting from about 2.4 to 2.1 bya. Recall that the sun was giving off much less heat at this time as well. A period when most of the Earth is glaciated and covered in ice is referred to as a Snowball Earth. Some think there was no Snowball Earth at all at this time; the idea is controversial. Also, some think depletion of methane by oxygen was not the cause of the glaciation, and it is almost certain it was not the sole cause. The Earth was certainly not entirely glaciated, because there was not enough water to form glaciers between 10° and 30° latitude, in both the northern and southern hemispheres. This allowed life to survive in these unglaciated areas.

Only prokaryotes were alive at the time of the Huronian glaciation and proposed Snowball Earth. They could have lived under the ice, deep underground, in pockets of liquid water under ice caps as occurs in Lake Vostok in Antarctica, in geothermal hot spots on land such as in Yellowstone National Park and places in Iceland today, in warmer waters in the deep sea, and at deep sea vents. Prokaryotes also can become dormant and live for hundreds of years in freezing temperatures, becoming active when the temperature increases. There could well have been some loss of diversity in the sea. However, microfossils show that life survived without any significant loss of diversity, and developed complex food webs, in spite of the glaciation, in shallow marine environments (Corsetti et al., 2003). If there were a mass extinction from a Snowball Earth, scientists would have observed a great reduction in biodiversity and a massive change in the species composition of Earth. They did not see this (ibid.). In addition, the groups that would have been most vulnerable to climatic shifts and extreme cold were not affected for the most part (Grey et al., 2003). Prokaryotes have short generation times; some can double their population in 20 minutes under ideal conditions. They typically have large populations. Thus, they are very adaptable and can evolve and adapt to environmental challenges such as extreme cold very quickly. Therefore, there almost surely was not a mass extinction as a result of the cold caused by photosynthesis.

When the Earth heated up and the glaciers thawed, nonbiological factors helped this, but life helped accelerate this. During millions of years, cryoconite, which is made of small rock particles, soot, *and microbes*, would have accumulated on and inside the ice, darkening it (Hoffman, 2016). Microorganisms that were adapted to extreme cold, as well as volcanic ash and dust from ice-free locations, would settle on ice, covering several million square kilometers of it. These would darken the ice, making it absorb more heat, accelerating the melting (ibid.). Thus, life in the form of microbes combined with nonbiological substances to help warm the planet and melt the ice. There were also other factors that helped with the thaw.

Significantly for the Pachamama Hypothesis, the Huronian glaciation may have helped the production of more oxygen in two ways. First, hydrogen peroxide, which has two hydrogen and two oxygen atoms per molecule of it, and is produced when ultraviolet light hits water, would have been protected from being broken down by sunlight when trapped in the glaciers. When the glaciers melted, it would be split into water and oxygen, releasing oxygen gradually but in great amounts (Liang et al., 2006).

Second, the melting of the glaciers when the glaciation ended released tremendous amounts of glacial deposit, resulting in sediments washing to the sea that were high in nutrients, such as phosphorus, causing phytoplankton blooms.

Life Coevolved with the Atmosphere

This chapter and the one before it demonstrate that life coevolved with Earth's atmosphere. Life reduced atmospheric carbon dioxide and methane levels, keeping acidity in aquatic ecosystems and temperatures low enough to favor life. Life increased the levels of oxygen in the air and seas, and it produced the ozone layer. These created an atmosphere that allowed life to increase in complexity, abundance, and diversity. The atmosphere allowed the evolution of complex land plants, including trees. This allowed life to further regulate greenhouse gasses and global temperatures, and to increase oxygen levels in the atmosphere, and in rivers, lakes, and the sea. These favorable carbon dioxide, methane, and oxygen levels, and favorable temperatures and levels of acidity in aquatic ecosystems, allowed life to further thrive, evolve complexity, and diversify, reaching spectacular levels of complexity and diversity. There was a constant interplay between life and the atmosphere. Life was instrumental in negative feedback loops that stabilized levels of carbon dioxide, methane, oxygen, fire, and temperature, keeping them all favorable to life. There were also positive feedback loops involving life and the atmosphere and sometimes other factors such as fire. However, in this case, they did not destabilize the system. Rather, the positive feedback sometimes even helped biology increase its biodiversity, complexity, and information content.

References

Kasting, J. F. (12 Feb. 1993). Earth's early atmosphere. *Science* 259 (5097): 920-6. doi: 10.1126/science.11536547

Sessions, A. L., Doughty, D. M., Welander, P. V., Summons, R. E., & Newman, D. K. (28 July 2009). The continuing puzzle of the Great Oxidation Event. Curr. *Biol.* 19 (14): R567-74. doi: 10.1016/j.cub.2009.05.054

Cardona, T. (March 2018). Early Archean origin of heterodimeric Photosystem I. *Heliyon* 4 (3): e00548. https://doi.org/10.1016/j.heliyon.2018.e00548

Halevy, S. E., et al. (20 July 2012). Sulfate burial constraints on the Phanerozoic sulfur cycle. *Science* 337 (6092): 331-4. doi: 10.1126/science.1220224

Catling, D. C., Zahnle, K. J., & McKay, C. P. (2001). Biogenic methane, hydrogen escape, and the irreversible oxidation of early Earth. *Science* 293: 839-43.

Pope, E. C., Bird, D. K., & Rosing, M. T. (2012). Isotope composition and volume of Earth's early oceans. *PNAS USA* 109: 4371-6.

Hoehler, T. M., Bebout, B. M., & Des Marais, D. J. (19 July 2001). The role of microbial mats in the production of reduced gases on the early Earth. *Nature* 412: 324-7. doi: 10.1038/35085554.

Hazen, R. M., et al. (2008). Mineral evolution. *Amer. Mineralogist* 93: 1693–720. doi: 10.2138/am.2008.2955 1693

Douglas, S. & Beveridge, T. J. (1 June 1998). Mineral formation by bacteria in natural microbial communities. *FEMS Microbiology Ecol.* 26 (2): 79–88. https://doi.org/10.1111/j.1574-6941.1998.tb00494.x

Granold, M., et al. (2 Jan., 2018). Modern diversification of the amino acid repertoire driven by oxygen. *PNAS USA* 115 (1) 41-6. doi: 10.1073/pnas.1717100115

Och, L. M. & Shields-Zhou, G. A. (Jan., 2012). The Neoproterozoic oxygenation event: Environmental perturbations and biogeochemical cycling. *Earth Science Reviews* 110 (1): 26-57. doi: 10.1016/j.earscirev.2011.09.004

Canfield, D. E., et al. (2014). *Oxygen: A Four Billion Year History.* Princeton Univ. Press, Princeton, NJ.

Shen, B., Dong, L., Xiao, S., Kowalewski, M. (2008). The Avalon explosion: Evolution of Ediacara morphospace. *Science.* 319 (5859): 81–84. doi:10.1126/science.1150279. Bibcode:2008Sci...319...81S. PMID 18174439. S2CID 206509488

Porada, P., et al. (Feb. 2014). Estimating impacts of lichens and bryophytes on global biogeochemical cycles. *Global Biogeochem. Cycles* 28 (2): 71-85. https://doi.org/10.1002/2013GB004705

Mills, D. B., et al. (18 March 2014). Oxygen requirements of the earliest animals. *PNAS USA* 111 (11): 4168-72. https://doi.org/10.1073/pnas.1400547111

Mills, D. B. & Canfield, D. E. (Dec. 2014). Oxygen and animal evolution: Did a rise of atmospheric oxygen "trigger" the origin of animals? *Bioessays* 36 (12): 1-11. https://doi.org/10.1002/bies.201400101

Logan, G. A., et al. (4 July 1995). Terminal Proterozoic reorganization of biogeochemical cycles. *Nature* 376 (6535): 53–6. doi: 10.1038/376053a0

Lenton, T. M. & Watson, A. J. (10 March 2004). Biotic enhancement of weathering, atmospheric O2 and carbon dioxide in the Neoproterozoic. *Geophys. Research Letters* 31 (5): L05202. Bibcode:2004GeoRL..31.5202L. doi 10.1029/2003GL018802

Heckman, D. S., et al. (10 Aug. 2021). Molecular evidence for the early colonization of land by fungi and plants. *Science* 293 (5532): 1129-33. doi: 10.1126/science.1061457

Beerling, D. (2007). *The Emerald Planet: How Plants Changed Earth's History.* Oxford Univ. Press, Oxford, UK; New York, NY. First two citations, page 14.

Lenton, T. M. (2012). Fires and the rise and regulation of atmospheric oxygen. *EGU General Assembly 2012, held 22-27 April, in Vienna, Austria.* p. 12930.

Lenton, T. M., et al. (30 Aug. 2016). Earliest land plants created modern levels of atmospheric oxygen. *PNAS USA* 113 (35): 9704-9 https://doi.org/10.1073/pnas. 1604787113

Dahl, et al. (19 Oct. 2010). Devonian rise in atmospheric oxygen correlated to the radiations of terrestrial plants and large predatory fish. *PNAS USA* 107 (42): 17911-5. doi:10.1073/pnas.1011287107

Berner, E. K., Berner, R. A., & Moulton, K. L. (2003). Plants and mineral weathering: present and past. *Treatise in Geochem.* 5: 169-88.

Falkowski, P. G., et al. (30 Sept. 2005). The rise of oxygen over the past 205 million years and the evolution of large placental mammals. *Science* 309 (5744): 2202-4. doi: 10.1126/science.1116047

Roche, J. W., et al. (Oct. 2019). Estimating evapotranspiration change due to forest treatment and fire at the basin scale in the Sierra Nevada, California. *Ecohydrology* 11 (7): e1978. doi: 10.1002/eco.1978

Barkley, A. E., et al. (29 July 2019). African biomass burning is a substantial source of phosphorus deposition to the Amazon, Tropical Atlantic Ocean, and Southern Ocean. *PNAS USA* 116 (33) 16216-21. https://doi.org/10.1073/pnas.1906091116

Tang, W., Llort, J., Weis, J., et al. (15 Sept. 2021). Widespread phytoplankton blooms triggered by 2019–2020 Australian wildfires. *Nature* 597: 370–5. https://doi.org/10.1038/s41586-021-03805-8

Kramer, S. J., Bisson, K. M., & Fischer, A. D. (20 Nov. 2020). Observations of phytoplankton community composition in the Santa Barbara Channel during the Thomas Fire. *Journ. of Geophysical Research: Oceans.* https://doi.org/10.1029/2020JC016851

Lenton, T. M. & Watson, A. J. (1 March 2000). Redfield revisited: 2. What regulates the oxygen content of the atmosphere? Global Biogeochem. *Cycles* 14 (1): 249-68. doi: 10.1029/1999GB900076

Hayward, M. W., et al. (20 March 2016). Could biodiversity loss have increased Australia's bushfire threat? *Animal Conservation* 19 (6): 490-7. https://doi.org/10.1111/acv.12269

Watson, A., Lovelock, J. E., & Margulis, L. (Dec. 1978). Methanogenesis, fires and the regulation of atmospheric oxygen. *Biosystems* 10 (4): 293-8. https://doi.org/10.1016/0303-2647(78)90012-6

Corsetti, F. A., et al. (7 April 2003). A complex microbiota from snowball Earth times: Microfossils from the Neoproterozoic Kingston Peak Formation, Death Valley, USA. *PNAS USA* 100 (8): 4399–404. doi: 10.1073/pnas.0730560100. Bibcode:2003PNAS..100.4399C. PMC 153566. PMID 12682298

Grey, K., Walter, M.R., & Calver, C.R. (1 May 2003). Neoproterozoic biotic diversification: Snowball Earth or aftermath of the Acraman impact? *Geology* 31 (5): 459–62. doi: 10.1130/0091-7613(2003)031<0459:NBDS EO>2.0.CO;2. Bibcode:2003Geo....31..459G

Hoffman, P. F. (2016). Cryoconite pans on Snowball Earth: supraglacial oases for Cryogenian eukaryotes? *Geobiol.* 14 (6): 531–42. doi: 10.1111/ gbi.12191. PMID 27422766

Liang, M.-C., et al. (12 Dec. 12 2006). Production of hydrogen peroxide in the atmosphere of a Snowball Earth and the origin of oxygenic photosynthesis. *PNAS USA* 103 (50): 18896-9. https://doi.org/10.1073/ pnas.0608839103

CHAPTER 5. SYMBIOSIS IS FUNDAMENTAL AND GREATLY INCREASES DIVERSITY

Symbiosis is Selected for, Very Common, and Greatly Increases Diversity

I am using symbiosis to mean a relationship between two species in which both benefit. An example is pollination, where the pollinator obtains a nutritious meal from the plant's flower and aids the plant by pollinating it, aiding it in reproduction. This is different from the other definition of symbiosis, which is a relationship in which one species benefits, and the other species either benefits (mutualism), or is unaffected (commensalism), or is harmed (parasitism). It is a corollary of the ABH that symbiosis is very common and important, helps structure ecosystems, and profoundly enhances diversity. This is because if, as the ABH states, all species make their ecosystem better for life and diversity, then it will be common that pairs of species help each other and that this relationship will often greatly enhance diversity.

Symbiosis clearly maintains species diversity by benefiting both species involved. I will show that it also greatly increases the number of species. I contend that its importance is underestimated, and that it is of crucial and fundamental importance in organizing and structuring ecosystems, affecting evolution, causing diversification, and increasing species diversity. The evolution of symbiotic interactions can result in macroevolution, with spectacular leaps by both species to new forms and into novel adaptive zones. When an adaptive zone is newly entered into, it allows for diversification into many new species. An adaptive zone may be looked at as a set of available niches. Examples of adaptive zones are the set of niches available when birds evolved flight or when fish evolved into amphibians and colonized land. Evolution into new adaptive zones gives access to many new niches and so is followed by diversification to many new species and forms. Thus, the coevolution of two species into a symbiotic relationship can lead to both species undergoing huge diversifications into many new species. These new species then provide habitat and niches which many additional species evolve to use; this second step causes a great amount of further diversification into numerous new species. Macroevolutionary leaps caused by symbiosis occurred at least ten times in the history of life. Symbiosis is a major driver of evolution.

Symbiosis is a stable state that evolution tends toward under the right conditions (there must be a fit where both species benefit). A large literature supports this. Symbiosis plays a fundamental role in all ecosystems (see Thompson, 2005). Doebell and Knowlton (1988) showed that, when increased investments in a different species yields increased returns, symbiosis can evolve with ease.

Attempts to determine how common symbiosis is indicate it is extremely common. Associations between plants and fungi, which generally are symbiotic (Lewis, 1973), are found in 83% of dicots, 79% of monocots (Trappe, 1987) (dicots and monocots are the two groups into which all flowering plants were formerly divided), and in *all* gymnosperms (conifers and their relatives; pine trees are gymnosperms) (Newman and Reddell 1987). The relationships between plant roots and mycorrhizal fungi may be as old as the evolution of land plants themselves (Pirozynski and Malloch, 1975; Pirozynski, 1981; Stubblefield et al., 1987). Symbiosis between plants and their animal pollinators are also highly diverse, common, and ancient (Crepet, 1983). And the number of plant species whose seeds are dispersed by animals and the number of animal species that disperse plant seeds are both huge.

Composite symbiotic organisms include the 2,500 coral species, of which 1,000 are hard coals that build reefs, promoting high diversity, and an estimated 13,500 to 17,000 species of lichens, extending from the tropics to the polar regions. All multicellular organisms are ecosystems harboring thousands of species of symbiotic and commensal viruses, bacteria, archaea, fungi, and other organisms. Even one-celled eukaryotes have a very diverse ecosystem of bacteria, archaea, and viruses.

Thus, symbiosis is economically adaptive, often selected for, often has a tendency to evolve, and is very common.

Following are some examples of symbiosis given to illustrate its ubiquity.

I have coined the term genetic coevolutionary symbiosis (Seaborg, 2022). This is symbiotic *evolution*. Species A helps species B as in any symbiosis. Species B helps the evolution of species A by giving it genetic material. One example of this is that viruses are commonly symbiotic with their hosts because they sometimes provide their hosts with new adaptive DNA segments. (Of course, viruses also aid their hosts in other ways, such as by regulating their populations by negative feedback). Genome-wide analyses have shown that eukaryotic genomes originated from a symbiotic coevolution between bacteria and archaea (William and Koonin, 2006, and references therein).

Microbial mats are multi-layered sheets of microbes, mostly bacteria and archaea, but also protozoa and other one-celled organisms. They are mainly in water or moist areas on land, but a few are in deserts, and a few are symbiotic in animals. Mats date to 3.5 bya, and are the oldest life on the planet represented by good fossil evidence. Since their first appearance up to the present, they have been communities of symbiosis, commensalism, competition, and predation. The fossil record of early photosynthesis shows that three distinct bacterial communities lived in vertical layers in mats along a gradient of oxygen availability, with each layer helping the other two layers, except the middle layer neither helps nor hurts the top one.

The symbioses and commensalisms in the mats were key drivers of many of the major evolutionary breakthroughs of life's history. Anoxygenic photosynthesis, which does not produce oxygen, likely evolved in mats. Oxygenic photosynthesis, which produces oxygen, probably evolved there too, as a result of the close proximity of different bacterial species, which brought together the two photosystems of oxygenic photosynthesis, perhaps with the aid of a virus. This led to life's production of oxygen, producing an atmosphere high in oxygen, allowing great diversification.

About 3.5% of prokaryotes are known to be involved in symbioses with other prokaryotes (Overmann & Schubert, 2002). But this has been studied little, so many more such symbioses likely await description.

The mitochondrion evolved when an archaeon ingested but did not digest a bacterium, and the two coevolved symbiotically until the bacterium became the organelle of the cell that carries out cellular respiration, using oxygen to create energy for the cell (Margulis & Sagan, 1986; Martin & Muller, 2007). This occurred about 2.2 bya. The mitochondrion gradually lost its capabilities other than respiration and become more and more dependent on its host cell for survival, while the host cell became more dependent on the mitochondrion for its energy. Cellular respiration by mitochondria is what allowed eukaryotic organisms to evolve into complex multicellular organisms. Without mitochondria, complex organisms would not be able to produce the energy needed to exist. The chloroplast, the cellular organelle that carries out photosynthesis for photosynthetic organisms such as plants and algae, evolved by the same process as the mitochondrion, starting about 1.6 bya. Here, an archaeon ingested a cyanobacterium that was capable of photosynthesis, and the two coevolved. Photosynthesis produces carbohydrate food and oxygen for the plant, algae, or other photosynthetic organism. Thus, *the eukaryotic cell evolved by coevolution and symbiosis.* This likely happened in microbial mats. *The close proximity of prokaryotic species in mats likely allowed the evolution of the chloroplast and mitochondrion by symbiosis, and hence the evolution of the eukaryotic cell.*

Molecular nitrogen in the air, N2, cannot be used by life. It must be converted to ammonia, NH3, which plants can use, by a process called nitrogen fixation. This process is necessary for higher life and is needed for ecosystems to function, because nitrogen is an essential element to life, being used in proteins, DNA, and RNA. Some bacteria, called nitrogen-fixing bacteria, convert molecular nitrogen to ammonia. Some of these bacteria live freely in the soil. But many are in structures called root nodules in the roots of plants. The plant provides them with a habitat, a safe place to live, while they convert nitrogen to a useable form for the plant. Thus, *symbiosis is fundamentally important in making the essential nutrient, nitrogen, available to life.*

Cornejo-Castillo et l. (2024) found a hitherto unknown symbiosis of an alga called *Braarudosphaera bigelowii* that engulfed a nitrogen-fixing cyanobacterium called UCYN-A about 100 mya. The two coevolved until the cyanobacterium became a part of the alga, and now fixes nitrogen for the alga in exchange for receiving a habitat from it. This is the same coevolutionary process by which the mitochondrion and chloroplast evolved to become organelles of cells. This organelle of the cell is called a nitroplast.

Bioluminescence is the production of light by organisms, causing them to glow. In most species, bioluminescence is produced by chemicals made by the glowing organism itself, but in some, the light is produced by symbiotic bacteria that live in their light organs (Hastings, 1978). The habitat where bioluminescence is most important and common is the deep sea, where light cannot penetrate. The need for light in this dark environment means that the symbiotic bacteria that provide it and the symbiotic relationship itself are key to structuring the entire deep-sea ecosystem and essential to its high biodiversity. There are 460 species in 21 families of marine teleost fish, the main fish family living today, that form symbiotic relationships with bioluminescent bacteria. Hundreds of species of invertebrates form such relationships. The deep sea displays spectacular light shows as a result.

The bacteria benefit because its host provides it with a mobile habitat, nutrients, and oxygen. The animals benefit by escaping predators through startling them; camouflage through counter-shading, whereby the animal's upper portion is dark and blends with the sea below it if viewed from above, while its belly is light due to the glowing bacteria and blends with the lighter area above it if viewed from below; prey warning predators of their unpalatability; misdirection of predators by such things as certain squid and small crustaceans expelling glowing bacteria; the prey making their predators easier for the latter's predators to see; locating, illuminating, attracting, startling, and confusing prey; warning and deterring competitors; recognition, attraction, and signaling of mates; and navigation in dark or low-light environments (Haddock et al., 2010).

Eukaryote-Eukaryote Symbioses

Lichens

Lichens are composed of a tight symbiosis, and are composite organisms, consisting of a species of fungus and a photosynthetic ally called a photobiont. Nearly 20% of the estimated 5.1 million fungal species are associated with lichens. Considering the number of symbiotic nitrogen-fixing fungi in plant roots, symbiotic leaf cutter ant fungi, and other symbiotic fungal species, the percentage of fungal species that are involved in symbiotic relationships must be well over 30% of all fungi. There are about 13,500–17,000 identified lichen species; there are certainly many more than this yet to be discovered. They range from the tropics to the poles.

At least 52 lichen genera on six continents have a third symbiont, a yeast, making the system at least a 3-way symbiosis (Spribille, 2016). The yeast likely produces chemicals that repel predators and microbial competitors, while getting a habitat from the fungus.

It is suspected that there is a fourth set of symbionts in lichens: bacterial communities (Grube et al., 2009). The lichen provides a habitat, protection, and nutrient to the bacteria. It is not certain, but the bacteria seem to be involved in nutrient cycling, hormone production, making nutrients such as phosphate available to the fungus and photobiomt, and fighting microorganisms that cause disease in the lichen.

The fact that there are often these two other symbionts and that nearly 20% of the estimated 5.1 million fungal species are lichens supports my thesis that symbiosis is very common.

Lichens are of tremendous importance in structuring ecosystems, and in the evolution and creation of biodiversity. It is estimated that 6% of Earth's land surface is covered by lichens. They increase biodiversity by living in environments where most life would not otherwise be able to live, including Arctic tundra, hot deserts, cooled lava flows, bare rocks, rocky coasts, exposed soil surfaces, tree bark, on tropical rainforest leaves, and toxic slag heaps. Several kinds of lichens grow on rocks, where they produce acids and other chemicals that break down rocks and other substances. Their root-like organs, called rhizines, penetrate rocks and help break them up. They are thus crucial in creating soil and contributing to soil quality. Lichens prepare the soil for more complex plants, and therefore animals, to take hold, on an ecological time scale. They also hold soil together, decreasing erosion by wind and water. Since cyanobacteria in some lichens fix nitrogen, some lichens fertilize the soil with nitrogen in a form available to life. Lichens absorb large amounts of water, and it is possible that a great many species of small animals, from insects and other invertebrates to small mammals, depend on or are aided by lichens for water, especially in winter. Thus, this *symbiotic, composite organism* aids and is in some cases symbiotic with many, many other species.

Lichens are networks of tightly inter-connected symbiotic species that can be among the largest organisms in the world, closely associated with trees, rocks, or soil. Coral reefs and higher organisms with their microbiomes are similarly networks. The fundamental unit of life may not be the organism, but the network.

Termites and Their Symbiotic Microbiome

Termites cannot digest the cellulose-containing wood they eat without the help of a complex microbiome in their gut. This symbiosis allowed them to exploit this adaptive zone, diversifying into about 3,106 described species, with an estimated few hundred more undescribed. The digestive system of the termite is a complex ecosystem with species from all three domains of life: bacteria, archaea, and eukaryotes. Many species in the termite gut are in symbiosis or commensalism with each other as well as the termite. Commensalism is a relationship between two species in which one benefits and the other is unaffected. An example is hermit crabs using the shells of dead snails. The microbes in the termite gut benefit because the termite provides them a habitat.

The termites and this symbiosis are beneficial to numerous other species. Termites decompose and recycle. They generally feed on fallen trees and stumps, leaf litter, and animal dung. They therefore decompose and recycle huge amounts of organic matter through the ecosystem. They aerate the soil. Groups of species such as drywood termites live in soil sometimes many feet deep. They build tunnels and carry out other activities underground, breaking down soil particles and increasing soil aeration. This allows plants to grow better. They add organic nutrients to the soil and concentrate them when they cart dead plant material and animal excrement to their underground nests. Termites help with water absorption in the soil.

Termite colonies provide good habitats with the temperature and humidity controlled and kept at favorable levels. They attract many species that utilize them. Kenya dwarf geckoes occur in high numbers near the mounds in Kenya. Golden-shouldered and hooded parrots nest in them in Australia. Various species of silverfish (a kind of wingless insect) and true bugs (order Hemiptera) use the mounds as habitat, and earthworms seek refuge in them when rain saturates the soil. Countless other species use their nests as habitat.

During droughts, rainforest termites show greater activity, resulting in greater soil moisture, partially countering the effects of the drought, and increasing nutrient mixing and plant-seedling survival (Davies et al., 2015).

Endophytes and Plants

Endophytes are endosymbionts, often bacteria or fungi, which use plants as habitats for at least part of their lives without causing disease. An endosymbiont is any organism that lives within the body or cells of another organism, usually, but not always, in a symbiotic relationship. They have been found in all plant species studied, and only a small percentage have been characterized. One leaf of a plant can have many bacterial and fungal endophyte species. Their relationships to their hosts are poorly understood, but they mostly greatly aid their plant hosts. They may benefit their hosts by preventing pathogenic organisms from invading them (Faeth, 2002). They do this by direct competition and production of chemicals that inhibit the growth of pathogenic competitors. Some bacterial endophytes increase plant growth. Although fungal endophytes can cause leaves to lose water at higher rates, certain of them help plants survive heat and drought. Endophytes helped the tremendous diversification of plants. There are several endophytes that inhabit seaweeds and algae.

Animal and Plant Symbioses, Sometimes Involving Other Taxa

There are many ways plants and animals are symbiotic. These often involve additional species symbiotic with them as well. The main examples of these are discussed below.

Leafcutter Ants, Plants, Fungi, and Bacteria

Leafcutter ants of 47 species (Speight et al., 1999) farm fungus. They occur only in South and Central America, Mexico, and parts of the southern United States. They are mainly in the rainforest. They cut small pieces of tree leaves, flowers, or grasses, depending on the ant species, and carry them to underground nests. The ants do not eat the leaves, but inoculate them with a species of fungus. They clean the leaves and seeds, removing anything that grows on them other than the fungus, eliminating its competitors, and pathogens of the fungus (Leal & Oliveira, 1998). If the ants unknowingly bring back leaves that are toxic to the fungus, it secretes a chemical that warns the ants not to gather any more of that species of leaf. The ants mulch and fertilize their crop, and their larvae eat the fungus, which is their only food (adults feed on the leaf sap). Thus, they literally farm the fungus, using the vegetation as food and a growth medium for the fungus. There clearly was coevolution between ant and fungus.

There are several nutrient-supplying bacterial species in a symbiotic relationship with the ants, and a symbiotic or commensal one with the fungus. The bacteria benefit from the ants and leaves by receiving food and a habitat. The bacteria break down sugars from the leaves, making a variety of nutrients that are likely used by the fungus and the ants. The ants may use the nutrients directly, or indirectly because they eat fungus that use the nutrients. The bacteria also transport sugars; make amino acids; and manufacture vitamin B5, which is needed to break down protein, carbohydrates, and fats, and to produce energy from nutrients. All of this benefits both their allies. It is not clear if the bacteria benefit from the fungus. They might get nutrients and/or habitat from it.

In another symbiosis, bacteria of the genus *Klebsiella* grow on the fungus that leafcutter ants grow, and fix nitrogen, which can be used by the fungus, ants, and likely other bacteria.

There is still another coevolved symbiosis. Leafcutters harbor antibiotic-producing bacteria of the genus *Pseudonocardia* that grow in their metapleural glands (secretory glands of ants) (Zhang et al., 2007). These bacteria keep the ants' leaf cuttings free of the parasitic sac fungus *Escovopsis*, which kills the fungus the ants eat. The symbiotic bacteria are nourished by special glandular secretions from the ants. The antibiotics are specifically targeted to attack *Escovopsis* (Currie et al., 1999).

Thus, this is a community of several species of symbionts, commensals, and prey (the plants). It is a web of many species of bacteria with the ant, plant, and fungus, and probably each species in the system directly or indirectly helps and is helped by every other species in the web, and at least most, and probably all, species co-evolved together. The ants help the plants by regulating their populations and dispersing their seeds.

Leafcutter ants are but one of about 250 ant species worldwide that farm and eat fungus, compared to a total of more than 12,500 described ant species and an estimated total of 22,000 if one counts species not yet described. Thus, the symbiosis resulted in a relatively small amount of diversification of ant species, although the total species that resulted from it is large, considering all the bacterial species that coevolved with the ants.

The ecological impact and increase in biodiversity on their ecosystem that resulted from this coevolved community symbiosis involving several symbionts is tremendous. Leafcutter colonies have over eight million individuals, and their farms are huge in area, yielding over a ton of fungus each year. The central mound of the underground nests can reach 98 feet across, and have smaller mounds radiating from it and extending to a diameter of 520 feet, occupying 320 to 650 square feet. They are estimated to account for 80% of all non-plant rain forest life! Nests of leafcutter ants of the genus *Atta* occur at up to six per hectare in tropical forests, especially in fragmented or edge forests.

Leafcutter ants are among the most conspicuous features of tropical forests. They are the rainforest's dominant herbivores, harvesting 10 to 50% of forest plants in their local area (Vasconcelos & Fowler, 1990; Wirth et al., 2003), harvesting leaves, twigs, bark, flowers, fruits, and seeds (ibid). They are important seed dispersers and predators, since they collect large quantities of seeds and fruits (Silva et al., 2009). The tremendous amount of seed collection is exemplified by the fact that the ants collect up to 160,000 seeds of the plant genus *Miconia* (family Melastomataceae) per day (Wirth et al., 2003)!

Their colossal system of tunnels augments soil porosity, aeration, water infiltration, and drainage (Cherrett, 1989). The ants create big soil mounds of up to 250 square meters (about 2691 square feet) (ibid), habitat for many microbial, plant, and invertebrate species.

The underground ants are a food source for a number of predators of the eggs, larvae, pupae, and adults. The adults also support a number of predators above ground.

Plants, Mycorrhizal Fungi, Animals, and Helper Bacteria

There is a four-way symbiosis between plants, animals that eat and bury nuts, fungi, and bacteria. Many plants, including conifers and flowering plants, have symbiotic fungi called mycorrhizal fungi associated with their roots. The fungus benefits because the plant gives it carbohydrates from photosynthesis, which the fungus cannot obtain on its own. The fungus provides the plant with a much greater root area to absorb more minerals and water. The fungal mycelium (plural mycelia), the non-reproductive part of it that grows underground, has a smaller diameter than the smallest root hair, so can penetrate soil that roots cannot, increasing surface area for absorption. The mycelia consist of numerous of branching, threadlike growing parts called hyphae, which are very long and increase the effective length of the plant's roots about tenfold. The fungi secrete acid that can dissolve or bond with charged chemicals called ions. Unaided roots may be unable to absorb ions that sometimes become immobilized, such as phosphate and iron. The fungus can obtain these for the plant. The fungi also actually make water from lignin and other chemicals. And fungal habitats are sponges that allow water to infiltrate the soil and stay there a long time, making it available to plants. The fungi break down rocks and make their minerals available to their plant allies.

The symbiosis has allowed a tremendous diversification of both plants and fungi. Mycorrhizal fungi are present in about 80% of land plant species and 92% of plant families studied, and are the most prevalent symbiosis in the plant kingdom (Harrison, 2005; Smith and Read, 2008). *About 90% of both flowering plant and conifer species have symbiotic mycorrhizal fungi with their roots* (Wang and Qui, 2006; for a review, see Smith and Read, 2008).

Amazingly, nutrients, carbon, and water can move between different plants through the mycorrhizal fungi (Simard et al., 2012), which form a large network of many interconnections, branches, and pathways, sometimes called the "wood wide web".

There are bacteria known as mycorrhizal helper bacteria, or MHB. They have been found every time they have been looked for, under many different environmental conditions, and in various plant-fungus symbioses; several MHB species have been found in the soil influenced by the fungi of many host plants (Andrade, et al., 1998-1; Andrade et al., 1998-2; Founoune et al., 2002; Garbaye, 1991). The evidence is clear that this represents a three-way symbiosis, that MHB help both root fungus and plant and are helped by both. Five mechanisms by which these bacteria help the plant and fungus were proposed by Garbaye (1994): (1) facilitating the root-fungal recognition process, (2) enhancement of fungal colonization of the plant, (3) helping the fungus obtain nutrients, (4) benefiting the surrounding soil, and (5) stimulation of germination of the fungus. Frey-Klett et al. (2007) presented clear evidence that the MHB help make nutrients from soil minerals available to the plant and fungus, aid with making atmospheric nitrogen available to the plant (that is, help with nitrogen fixation), and protect plants against bacteria that cause disease to their roots. The plant root and fungus benefit the MHB by providing them a home.

The plant-fungus-bacteria complex together perform a crucially important function to the ecosystem. This complex solubilizes minerals in rocks, such as phosphorus and iron, changing them from unobtainable to available to the plants, fungi, and bacteria, and life in general. Weathering and solubilization by this complex of organisms are the major sources of plant nutrients, except nitrogen, in natural ecosystems.

There is a fourth symbiotic ally: animals. Various species of deer, wild boar, mice, flying squirrels, other tree squirrels, and others eat the fruiting bodies of various fungus species. The fruiting body of a fungus has the spores with which it reproduces; some fruiting bodies are mushrooms. The animal then moves around and defecates the spores of the fungus, dispersing them. Sometimes the animal even buries the spores, placing them closer to plant roots. *The animal inoculates the plant root with the fungus.*

It is now apparent plants could not have colonized land without help from fungi. Since land plants evolved from aquatic green algae, many of the earliest land plants lacked true roots. Instead, they had rhizoids, root-like structures that anchored them to the soil. They could not have obtained the soil nutrients they needed to survive without their fungal allies. Of plants living today, plants with poorly developed root systems are much more responsive to mycorrhizal fungi than plants with complex, well-developed, branching root systems (Read et al., 2000). One of the biggest challenges plants faced since their appearance on land has always been extracting water and nutrients from the soil, which mycorrhizal fungi greatly help them to do. Direct fossil evidence supports this as well. Fossilized plants dating from 400 mya show well-preserved structures that look identical to modern fungal hyphae that penetrate root cells and form highly branched structures (Remy et al 1994). The earliest land plant fossils are 450 million years old, and genetic evidence based on a molecular clock indicates that the mycorrhizal fungi originated between 353 and 462 mya (Simon et al., 1993). Fungi helped build the soil for plants. *Tortotubus*, a 440-million-year-old genus of fungus, recently found in fossilized form, created rich, deep soil, and helped incorporate oxygen and nitrogen into the soil. Its ability to store and transport nutrients by decomposing dead organisms helped create a nutrient-rich layer of topsoil that was necessary for plant life to colonize land (Smith, 2016). In fact, Pirozynski and Malloch (1975) hypothesized that terrestrial plants are the product of an ancient and continuing symbiosis of a semi-aquatic ancestral green alga and an aquatic fungus. The colonization of land by plants and indeed the very evolution of plants, was possible only through such symbioses, which allowed plants to obtain water and nutrients (ibid).

Symbioses Exclusively Between Animals and Photosynthesizers

Plants and animals are in a profound symbiotic chemical relationship. Photosynthesis is the reverse chemical reaction to respiration. Through photosynthesis, plants use water and CO_2 and the energy of sunlight to produce carbohydrates and oxygen. Animal eat the carbohydrates and use them as a food source, and breathe the oxygen. Through respiration, animals (and plants) use oxygen and carbohydrates to produce CO_2, water, and energy. The energy produced from respiration is necessary for all higher life. This chemical symbiosis is essential for animals to survive, and greatly helpful to plants. One might consider it a fortuitous mystery that photosynthesis and respiration are opposite, complimentary chemical reactions.

Plants, phytoplankton, and other photosynthesizers are the basis of the food chain supporting all animals, directly or indirectly. Plants also provide animals with habitat, shelter, shade, and protection from predators. They regulate global temperature and acidity through the burial of carbon, regulating levels of the major greenhouse gases, CO_2 and methane, keeping their atmospheric levels beneficial to life. They regulate local climate and moisture by producing rain and providing shade. They stabilize the soil, give homes to bacteria that make nitrogen available to life, and provide other services to animals. Animals regulate plant populations to the benefit of plants; till, aerate, fertilize, and enhance the soil; pollinate; disperse seeds; and provide other services to plants. This general symbiotic relationship between the many species in these two kingdoms includes many examples of coevolution that benefited both kingdoms, resulting in a phenomenal increase in biodiversity in both taxa.

Two of these major symbioses, pollination and seed dispersal, are discussed below.

Pollination and Seed Dispersal of Plants by Animals

Angiosperm is the scientific term for flowering plants. Pollination and seed dispersal by animals are fundamentally important to angiosperms, and play a crucial role in the maintenance of forest ecosystems throughout the world, while simultaneously benefiting animals. Animals as pollinators and seed dispersers have long been suspected as agents of angiosperm diversification (Crepet, 1984; Regal, 1977; Stebbins, 1981), and theory indicates these could drive plant diversification by both increasing speciation (McPeek, 1996) and lowering extinction (Koh et al., 2004). It is now clear that the spectacular evolutionary success of tropical angiosperms over the past 145 million years is attributable to their reliance primarily on animals, rather than the wind, for pollination and seed dispersal. The diversification of angiosperms is one of the greatest on land in Earth's history, and has profoundly and fundamentally altered terrestrial ecosystems. *The 250,000 to 400,000 species of angiosperms make up about 90% of all living plant species* (Thorne, 2002; Scotland and Wortley, 2003; Govaerts, 2003). *Essentially all angiosperm species are pollinated and have their seeds dispersed by animals, which is at least two symbioses per angiosperm species, showing the ubiquity of symbiosis.* There is good evidence showing insect pollinators greatly promoted early angiosperm diversification (Crepet, 1983). It has been proposed animal seed dispersal led to the later Tertiary Period angio-sperm radiations (Crane et al., 1995). The Tertiary Period was from about 66 to about 2.6 mya. Pollination and seed dispersal diversified animals tre-mendously as well.

Pollination of Angiosperms by Animals

Pollination is symbiosis because the plant benefits by being pollinated, which is necessary for reproduction, and the pollinator receives a nutritious meal of pollen, nectar, or both. Animals and angiosperms have coevolved in this symbiotic relationship. It is believed flowers arose as a result of coevo-lution with animals, because flowers are designed and specifically adapted for animal pollination, and there is no other reasonable explanation for the evolution of the flower. The only purpose of a flower is to attract animal pollinators. Animal pollination is more efficient than wind pollination. It is an efficient agent of cross-pollination, which prevents self-fertilization, thereby maintaining genetic diversity and healthy offspring.

The origin of flowers was largely orchestrated by insects. Crepet (1983), in reviewing the literature on angiosperm evolution, suggested that there were two very rapid periods of flowering plant evolution. The first was associated with the flower's origin, driven by insects as the selective agents.

The second period was adaptive radiations of flower forms and pollina-tion mechanisms driven by and associated with the diversification of insect forms. The flower was a key adaptation that allowed more efficient pollina-tion and allowed flowering plants to enter into a new adaptive zone with many new available niches, resulting in tremendous plant diversification. As mentioned, angiosperms are the most diverse land plants today. This major evolutionary transition and diversification happened via coevolution with insects. Animal pollinators have increased angiosperms diversity spec-tacularly, and vice versa, each by acting as a coevolutionary selective force on the other.

Pollinators have caused plants to evolve a great diversity of species and floral types (Dafni et al., 2005, and references therein). For example, bee-pollinated flowers are yellow and blue, with ultraviolet nectar guides, since bees see these colors, but not red. Moths are nocturnal and have a good sense of smell, so the flowers they visit are white or pale to show up in moonlight, often only open at night, and have a strong, sweet perfume that advertises their presence in the dark.

Baker and Hurd (1968) pointed out that the vast majority of flowering plants are insect-pollinated. Ollerton et al. (2011) estimated the number and proportion of flowering plants that are pollinated by animals using published and unpublished community-level surveys of plant pollination systems that recorded whether each species present was pollinated by animals or wind, finding animals pollinated an average of 78% in temperate-zone communities and 94% in tropical ecosystems. By correcting for latitudinal diversity trends, they estimate that worldwide, about 308,000 angiosperm species, or 87.5%, are animal-pollinated.

It is estimated that there are about 200,000 animal pollinator species (Landry, 2010). This immense number of animal pollinators gives some indication of how common symbiosis is in animals, and how much coevolution with the plants they pollinate has increased animal species diversity; this is especially true with insects.

The insect orders that have the highest percentage of species that are pollinators also have the most total number of species. This includes beetles (order Coleoptera), which have a great percentage of pollinators, and about total 400,000 total species (Hammond, 1992), having more species than any other order of any multicellular organism, constituting almost 25% of all known animal species, and about 40% of all described insect species, with new species constantly being found. Bees, wasps, and ants (order Hymenoptera) have a high percentage of pollinating species, and 150,000 recognized species, and many more to be discovered. Butterflies and moths (order Lepidoptera) seem to have the highest percentage of pollinating species of any order; they have a total of 180,000 described species in 126 families (Capinera, 2008) and 46 superfamilies (Mallet, 2007), making up about 10% of all species of organism (ibid). Flies and mosquitoes (order Diptera) have about 150,000 described species and likely over 250,000 total species in 150 families, and many flies and mosquitoes are pollinators.

When symbiosis between two groups coevolves in some cases, such as pollinatio, it leads to tremendous diversification into great numbers of species in both groups. In the case of flowering plants and beetles, the first pollinators to evolve, this resulted in great diversification into myriad flowering plant and beetle species. This can then lead to other groups of organisms coevolving into carrying out the same symbiosis. For example, once flowers appeared from coevolution with beetles, many species of butterfly, moth, fly, mosquito, bee, and wasp coevolved as pollinators with flowers, producing many new species in these insect groups and many new species of flowers. And several new pollinator species of bird, bat, primate, and so on coevolved along with the many new flowering plant species that they became pollinators of. After the new pollinating species and their flowering plants evolve, both new pollinator and new plant species greatly diversify into several species. After the pollinators and pollinated plant species have coevolved and become established, another step of coevolution happens when other species evolve that benefit from the many new coevolved species of flowering plants and their pollinators. These include predators, herbivores, and parasites of the new pollinators and plants, species that use them as habitat, and so on. These new beneficiaries also coevolve with their new benefactors and diversify. This process of diversification of the symbiotic groups into many species and then the evolution of many new beneficiary species that benefit from the new symbiotic species occurred in many coevolved symbioses, including the mitochondrion, chloroplast, seed dispersal, symbiotic root fungi, lichens, coral, and others.

Seed Dispersal by Animals

Many kinds of animals disperse plant seeds (Van der Pijl, 1982, and references therein). Animals are key dispersers of seeds in tropical forests, which are notably dependent on fruit-eating birds and mammals for dispersal of their seeds.

Seed dispersal by animals helps plant species in at least seven ways. (1) It prevents the offspring from being shaded out by its parent. (2) It reduces the competition between offspring, and between the parent plant and offspring, for nutrients, water, and sunlight, which they would experience if they germinated near the parent plant. (3) It increases the probability of cross fertilization, preventing inbreeding, thus keeping the variability of the population high. (4) It increases the probability that seeds and seedlings will escape such natural enemies as invertebrate and vertebrate seedeaters, herbivores, and pathogens. (5) Widespread dispersal by mobile animals such as birds and bats allows plants to colonize new habitats. (6) Animals often selectively drop seeds in areas favorable to survival of the plant. (7) Seeds surrounded by fleshy fruits that go through animal digestive tracts often have greater germination percentages and speeds, and germinate at different times, decreasing seedling competition. All these factors increase seedling survival rates. Although the first five advantages of seed dispersal are also conferred by wind dispersal, the last two are unique to animal dispersal, and animals can be more effective than wind at some or all of the first five.

Seed dispersal resulted from coevolution, for plants coevolved tasty and nutritious fruits that attract animals, giving them an incentive to eat them. Animals evolved a taste for fruits and color vision that helps locate fruits. Thus, the animal benefits by getting a tasty, nutritious meal. Animals often disperse seeds far from the parent plants as a result of eating the fruit, moving away, and then spitting out or defecating the seeds. The seeds of most tree species are animal-dispersed. An estimated 51 to 98% of canopy and sub-canopy trees in New World forests and 46 to 80% in Old World forests are vertebrate-dispersed (Stoner and Henry, 2009).

Tropical forests are dependent on fruit-eating birds and mammals for dispersal of their seeds, and could not exist without them. At least 300 plant species in the Old World are known to rely on fruit bats for their propagation, from pollination, seed dispersal, or both. Bats are fruit-dispersers of major importance in tropical forests, essential to keeping the system diverse, healthy, and functioning. Birds are important seed dispersers. Fruits have evolved many specialized enticements for birds.

In the Amazon River system, many species of fish eat the fruit that drops into the river. Some fish species migrate to the upper reaches of streams in the Amazon River system, where fruits grow more densely. Here, the fruits start to ripen in the rainy season when the water is high and the forest flooded, so fish have the best access to them. The fish and trees thus co-evolved for fruit dispersal by fish. Worldwide, more than 100 different fish species have been found with viable seeds in their guts.

Bello et al. (2015) found hunting by humans decreased hardwood tree diversity and abundance in 31 tropical Atlantic forests in Brazil, with larger trees being replaced by smaller species, resulting in a 10 to 15% loss in carbon sequestration. This is significant, since tropical forests are responsible for about 40% of carbon sequestration worldwide. The reason for the tree loss is the decrease in seed dispersers.

Plants are part of a large coevolved, symbiotic, super-organism-like, very tightly interacting network consisting of: chloroplasts and mitochondria that are the coevolved evolutionary remnants of prokaryotes; endophytes; mycorrhizal fungi, helper bacteria, and animals that inoculate plant roots with this fungus; pollinators; seed dispersers; other plant species they communicate with; symbiotic soil organisms; predators of their herbivores; epiphytes; and even their herbivores. Their key roles in oxygen production, carbon sequestration, albedo regulation, stabilization of the water cycle, climate regulation, soil retention, river formation, and other such services to the ecosystem that in turn supports them are further manifestations of this coevolved, symbiotic, interacting network.

Coral Reefs

Coral is a composite of two species living in intimate symbiosis, as if they were one organism. The coral polyps are animals in the phylum Cnidaria, the phylum of jellyfish. Their symbiotic allies are photosynthetic algae called zooxanthellae that live in their tissues, providing 90% of the organic nutrients that nourish the polyp (Spaulding et al., 2001), including amino acids that build proteins, and glucose and glycerol for energy. They also supply up to 90% of the polyps' energy requirements (Sherman, 2009; Marshall & Schuttenberg, 2006). Without zooxanthellae, coral would grow too slowly to form a reef. The polyps secrete hard exoskeletons of calcium carbonate that support and protect the algae and themselves, and form the coral reef. Polyps also provide a constant supply of the CO_2 the algae need for photosynthesis. Also, polyps obtain nitrogen from zooplankton, and share some of it with the algae, which need it. The coral reef is an ecosystem of incredible biodiversity, having the highest number of species of any aquatic ecosystem, marine or freshwater, and higher species diversity than any terrestrial ecosystem, with the possible exception of tropical rainforests. There are a huge number of coral species, each polyp being associated with its specific alga species, and each alga with a unique polyp species. The hard, carbonate reef provides a three-dimensional habitat for a tremendous diversity of animals. Although coral reefs cover less than 0.1% of the world's oceans, they support over 25% of all marine species (Spalding & Grenfell, 1997). Over 4,000 species of fish inhabit coral reefs. The amazing diversity includes great numbers of species of seabirds, worms, crustaceans, mollusks (including clams, squid, and octopuses), jellyfish, sponges, sea squirts, echinoderms (including sea stars, sea cucumbers, and sea urchins); a few species of sea snakes and sea turtles; and even visits by whales and dolphins. Inside coral are a great diversity of crabs, shrimps, snails, and other invertebrates.

There is a spectacular amount of symbiosis and commensalism on coral reefs. Some of these directly involve coral and hence the reef structure. Many species of sponges are symbiotic with coral, and necessary for reef survival. Coral provides them a habitat; they live in crevices in the reef. Sponges are among the most efficient of any filter feeders of phytoplankton. They consume about 60% of the phytoplankton that drifts by in the Red Sea. Some reef nutrients exist in a form coral cannot use. Some species of sponges filter these nutrients, converting them to small particles that can be utilized by coral (Kaplan, 2009), and by algae. Yet these and other sponges could over-run the reef if not controlled. This is prevented because sea slugs, sea anemones, and sea turtles, mainly hawksbill turtles, eat and control sponges. Seaweed is controlled and kept at a level beneficial to the reef by its herbivores, including certain species of sea urchins and sea slugs. Otherwise, it would over grow and suffocate the coral. At intermediate levels, it benefits reef animals as a food source, by removing CO_2, and by producing oxygen. Foliose macroalgae would inhibit and stunt coral growth and even greatly reduce coral and reverse its development, as well as negatively impact the reef's resilience, and other algae would over-grow and suffocate the reef, but for herbivory of them by certain species of parrotfish and some sea urchins, such as the long-spined sea urchin (*Diademed antillarum*), which lives in holes in the reef. These herbivores ensure the algae is maintained at a more or less optimal level for reef health and diversity. Without them, the reef system would be destroyed. All relationships discussed in this paragraph are symbiotic; for example, sea urchins and sponges benefit from coral as habitat while helping coral.

Corals in Mo'orea, French Polynesia, provide a home to several species of coral guard-crabs, genus *Trapezia*. This coral is a favored food of the large crown-of-thorns sea star, which could devastate the reef, but for the protection by this large crab, which defends its home by pinching, shaking, and nipping the sea star's tube feet, which are small, flexible, hollow appendages that occur in great numbers on each of the sea star's five legs, used for movement and collecting food. The defense is effective. Intermediate-sized and small crabs of this genus defend the coral against predators of their size classes. Crabs do not attack predators that do not match their size class. Small crabs such as Serene's guard crab (*T. serenei*) and *T. punctimanus* (no common name) defend against small sea snails such as the horn drupe (*Drupella cornus*), which large crabs ignore, and medium-sized crabs, including the rusty guard carab (*T. bidentata*) and the slightly larger Serene's guard crab protect from intermediate-sized cushion sea stars (*Culcita novaeguineae*). The crabs also protected smaller coral species that lived on the large coral, and this greatly aided recovery from massive crown-of-thorns predation when this species had a population explosion due to human effects. The coral offer shelter and produce fatty deposits in their tips that the crabs eat (McKeon and Moore (2014), showing coevolution in this symbiosis.

Müllerian Mimicry

Müllerian mimicry is symbiosis where two distasteful, poisonous, or venomous prey species benefit by resembling each other, allowing their predator to learn to avoid eating either of the two species. It is common in butterflies. Unpalatable *Heliconius* butterflies in Central and South America, the postman butterfly (*H. Melpomene*) and the red postman (*H. erato*), have geographic races that differ from other races of their own species, but resemble races of the other species that live in the same area with them and are also unpalatable. In Peru, these species are involved in a mimicry ring with other *Heliconius* species, three other genera, and a moth; all these species are unpalatable, being full of bad-tasting toxins (Mallet, 1999; Futuyma, 2005). These colorful insects are illustrated in Figure 5.1. There are a number of other cases of mimicry rings, especially in butterflies. Müllerian mimicry is widespread and common, especially in insects.

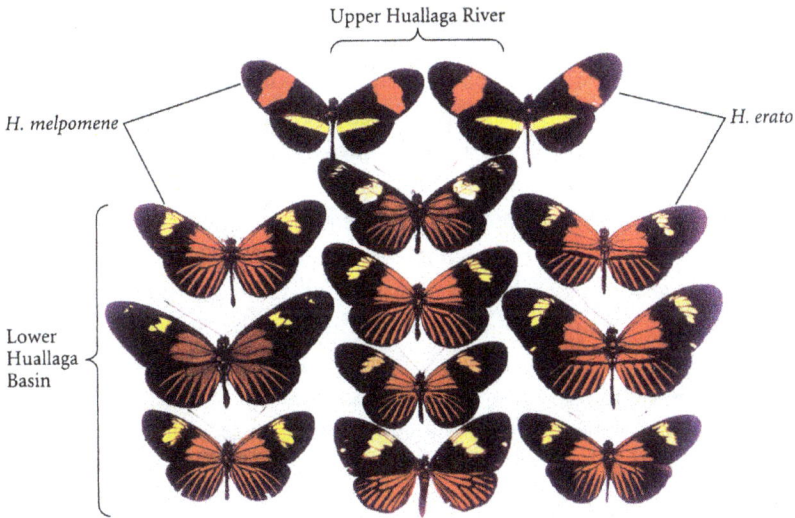

Figure 5.1. A mimicry ring consistent with the ubiquity of symbiosis in nature. The postman butterfly (*Heliconius melpomene*) and the red postman (*H. erato*) have a very different color pattern in the Mayo and upper Huallaga Rivers, in eastern Peru, than in the lower Huallaga drainage, where they join a mimicry ring with a "rayed" wing pattern. This ring of toxic, unpalatable species includes four other species of *Heliconius*, three other genera of butterflies (the top three species in the center column), and a moth (center column, bottom). From Mallet, 1999; Futuyma, 2005.

Chapter Five

Additional Examples of Symbiosis in Nature

Hydrothermal vents, which are fissures on the seabed from which geo-thermally heated water discharges, exist where no light penetrates. So photosynthesis cannot occur. Thus, heat and chemical synthesis (the construction of complex chemical compounds from simpler ones) by bacteria and archaea provide the energy to drive the system. Some archaea can live at temperatures of 2350 F. The bacteria and archaea use hydrogen sulfide and methane, both poisonous to oxygen-breathing animals, to produce the energy that is the primary production that runs the system. There are many symbioses between microbes and multicellular animals in which the animals supply a home to the microbes, while the microbes produce nutrients for the animals. Strange, about 6-foot-long, white tube worms, with gills that look like red plumes, called Pompeii worms (*Alvinella pompejana*), with no mouths or stomachs as adults, live in tubes on the sides of scalding black-smoker chimneys. It is believed that the worms staying near the sulfur-rich vent fluids allows them to provide nutrients for bacteria that carry out chemosynthesis to grow on them. The bodies of the tube worms serve as habitat for the bacteria. They have mouths when very young, when they take in the sulfur-eating bacteria. Their mouths disappear as they grow. They absorb oxygen from seawater and hydrogen sulfide from vent fluids, which feed the bacteria inside them. Their bacteria feed them with chemicals made by chemosynthesis. Other species of similar tube worms with similar symbioses with internal microbes are at other hydrothermal vents. Worms at the vents off the Galapagos Islands called thick-plumed rift worms (*Riftia pachyptila*) are an example. The gills of giant clams and mussels that live at the hydrothermal vents are full of and provide homes for symbiotic bacteria that provide them with nutrients, allowing them to grow very rapidly and very big. Since the discovery of hydrothermal vents in 1977, scientists have found more than 500 organisms that had never been seen before, all potentially symbiotic with microbes.

There are dinoflagellates that are symbiotic with clams, photosynthesizing for them, while receiving a habitat in which to live in return.

Certain species of flat worms, some arrow worms, some moon snails, three or four species of the venomous blue-ringed octopus, pufferfish, the Western newt, and some toads produce the very potent poison, tetrodotoxin, sometimes in combination with other poisons. It is used for defense, or, in the octopus, primarily to kill prey and secondarily for defense. The toxin is produced by symbiotic bacteria of six genera (Lago et al., 2015).

Various ant species in Australia obtain sugar water from the glands of caterpillars of gossamer-winged butterflies, the family which includes the blues. The caterpillars make a sound that signals to the ants that the nutrient is coming. It is thought the caterpillar secretes appeasement chemicals that cause the ant not to view it as prey. The ants protect the caterpillar from insect predators by fighting them off, and from bird predators because they often cover the caterpillar, and are distasteful to birds.

The Hawaiian shrimp goby, a fish, looks out for danger for either of two species of snapping shrimp (*Alpheus rapax* and *A. rapacida*), which are almost blind. The shrimp constructs and maintains burrows in the seabed, while the fish stands guard. The fish wiggles its tail against the shrimp's long antennae at the burrow entrance to warn the shrimp if a predator approaches. Then the shrimp goes into its burrow for safety. During construction, shrimps leave the burrow to deposit excavated sand. Shrimps use their antennae to maintain constant contact with the goby. Sometimes the goby even hovers above the shrimp, allowing it to take its load further from the burrow's entrance. In return, the goby uses the shrimp's burrow as a home, sleeping in it at night, and bolting into it to escape predators.

Cleaner fish remove dead skin, external parasites, and infected tissue from the skin and gill chambers of larger fish species. They also clean sea turtles, marine iguanas, manatees, whales, and octopuses. Wrasse, cichlids, catfish, pipefish, lumpsuckers, and gobies act as cleaner fish across the globe in fresh, brackish, and marine waters, but specifically concentrated in the tropics due to the high parasite density there. There are also cleaner shrimps.

American badgers and coyotes work as a team in hunting. Badgers dig up burrows of rodents such as prairie dogs and ground squirrels to catch and eat them. Coyotes stand nearby, and often catch and eat the rodent when it runs out of an alternate exit to escape the badger. Coyotes catch an estimated one third more rodents when teaming up with badgers than when hunting alone. Some biologists think the badger does not benefit from this relationship. But it almost certainly does, since the coyote sometimes scares the rodent back into its burrow, where the badger can dig quickly and capture it. And the presence of the coyote causes the rodent to tend to stay underground, where the badger has a better chance of locating it.

Common ravens (*Corvus corax*) associate with gray wolves (*Canis lupus*) in winter in Alaska and Canada as a symbiotic foraging strategy (Heinrich et al., 2002). They circle above prey, such as caribou and other species, signaling the prey's location to wolves, which have learned to look for circling ravens. Ravens stand sentinel for wolves while wolves eat, are more nervous and alert than wolves at kill sites, give them added vision and hearing, and alert wolves to danger, including animals that might steal their food. Ravens grab parts of the carcass as wolves are eating, and finish the carcass after the wolves are done. They also lead wolves to carcasses too tough for the ravens to break into and eat without the wolves opening them up first. Wolves howl before they go on a hunt, and ravens fly toward the wolves when they hear this signal. Wolves may respond to certain raven vocalizations and behavior that indicate prey is nearby. Ravens peck at wolves' behinds, pull their tales, and fly just out of reach, causing the wolves to chase them, in a game that reinforces their bond. The relationship is also competition, for once wolves kill the prey, ravens scavenge and steal up to a third of it from the wolves. Studies show the once commonly-held belief that wolves are in packs to hunt more efficiently is wrong, because meat consumed per wolf decreases as pack size increases in the absence of scavengers, since pack animals have to share meat (Vucetich et al., 2003). But if scavengers are present, meat per wolf increases with pack size, because packs better defend the kill from scavengers, the main one of which is the raven, at least in the Pacific Northwest. Thus, forming packs and sociality evolved in wolves as a result of selection pressure from ravens, in an example of one species influencing another's evolution. Only with packs and social behavior could wolves have become pre-adapted to domestication into dogs. Hence without ravens, humans would not have dogs! Human evolution would also have been quite different had we not evolved symbiotically with wolves, aiding each other in hunting. Humans shared food with wolves and eventually domesticated them as dogs. Remarkably, when wolves were re-introduced to Yellowstone National Park after having been absent from there since the 1940's, the ravens and wolves seemed to remember their relationship, and it was quickly re-established.

Nitrogen is a limiting nutrient in most terrestrial ecosystems, and had been thought to be supplied only by free-living bacteria and bacteria in plant root nodules. But Nardi et al. (2002) have found evidence that symbiotic nitrogen-fixing microbes of diverse forms are widespread in the hindguts of arthropods, with nitrogen-fixing rates as high as 10-40 kg/hectare/year being possible. Arthropods are the phylum that includes insects, spiders, and crabs.

Several species of *Acacia*, a genus of trees that are symbiotic with ant species, most notably in Mexico, Central and South America, and Africa. There are many species of ants that share in this symbiosis, and each *Acacia* species tends to ally with only one ant species. Some *Acacia* species have small, reinforced structures that the ants hollow out and use for nests. *Acacias* feed ants with nectaries, small stores of sugar-rich nectar on their stems that are not within their flowers, and nutritious structures called beltian bodies on their leaf tips. The ants guard their *Acacia* home, attacking any animal that may try to feed on it. *Acacias* have evolved mechanisms to prevent the ants from attacking their pollinators. *Acacia* flowers produce volatile compounds that are repellent to ants, causing them to stay away from the flowers (Raine et al., 2002). Symbiotic bacteria colonizing the ants inhibit the growth of disease-causing bacteria on the plants' leaves, and increase plant health (González-Teuber, et al. 2014). The increase in ant, *Acacia*, and bacterial diversity as a result of this symbiosis is very high because it has led to great diversification in all of these groups.

This is but a small percentage of the many symbioses in nature. They illustrate that it is common and important in increasing biodiversity.

When One Considers Indirect Symbiosis, Symbiosis is Very Common

If one considers indirect symbiosis, symbiosis is much more common than one would imagine. Indirect symbiosis is symbiosis in which two species help each other through an intermediary species. For example, let us say that species A and species B help other (are symbiotic), and species B and species C help each other. Thus, species A and species C help each other indirectly and are indirectly symbiotic. Any two species separated by two levels on a food web, such as a predator and a plant, are indirectly symbiotic, because the predator controls the population of herbivore that eats the plant and the plant feeds the predator's prey. Earthworms help numerous plant species by improving the soil. These plants help many herbivores species. Some of these herbivores are eaten by predators that also eat the earthworm's predators. This is a network of indirect symbioses. Similar symbiotic networks exist in all complex ecosystems.

Ecosystems Help Other Ecosystems that are Sometimes Far Away from Them. This is Often Carried out by Life

Ecosystems often provide services such as nutrients or protection to other ecosystems, in "commensalism between ecosystems". Sometimes two ecosystems aid each other, in "symbiosis between ecosystems". I discuss both types together here, rather than putting the commensal ones in the commensalism chapter, which follows this one.

Nutrients are carried from the photic zone, which is the part of the sea where enough light penetrates for photosynthesis to occur, to the deep sea by marine snow, sinking feces, and sinking dead organisms. Marine snow is a constant rain of falling organic matter in the ocean that mostly comes from organisms, and consists of dead or dying animals and plankton, other one-celled organisms, fecal matter, sand, and inorganic dust. Viruses of bacteria in deep sea sediments liberate nutrients, many of which come from near the surface, for use there, by attacking prokaryotes. A great portion of these nutrients are transported to the to near-surface waters by animals, contributing to photic zone ecosystems. This way, the area near the sea's surface and the seafloor exchange nutrients, aiding each other.

Many species of squid migrate from 500 to 1,000 meters (about 1,640 to 3,281 feet) deep, in the day, to within 150 meters (about 492 feet) of the surface to feed at night, transferring nutrients from deep to shallow waters if eaten near the surface, and in the other direction if consumed in the depths. Most transfers are from deep to shallow. Cephalopods (the class including squids and octopuses) have migrated in this manner for at least 150 million years.

Coral reefs on the one hand and mangrove and sea grass ecosystems on the other display symbiosis between ecosystems. Reefs that lie close to the shore physically protect mangroves and sea grass from waves and currents, and produce sediment in which the mangroves and sea grass can take root. Reefs protect land ecosystems from erosion, especially wetlands and islands, notably the islands between India and Sri Lanka. Mangroves and sea grass protect coral reefs from influxes of fresh water, which harm these saltwater ecosystems, and from large influxes of silt and pollutants. They both supply reefs with nutrients, and provide food and shelter as nurseries for the larvae and juveniles of many reef fish, with the subadults and adults migrating from these ecosystems to reefs. This is true in the Caribbean (Dorenbosch et al., 2004) and the Indo-Pacific (ibid.). Mangroves also provide nurseries for the larvae of coral and other reef invertebrates.

In North America, the grizzly bears transfer nutrients from the sea to the forest. Salmon, rich in nitrogen, sulfur, carbon, and phosphorus, swim up rivers, sometimes for hundreds of miles. The bears capture the salmon to eat them, carrying them onto dry land, defecating and dispersing partially-eaten fish. It has been estimated that the bears leave up to half of the salmon they harvest on the forest floor. This benefits a huge number of species of land plants (including trees), which benefit great numbers of animals. Bald eagles and osprey transfer vast amounts of nutrients from freshwater rivers and lakes to land through fish consumption and defecation.

Symbiosis is ubiquitous and common in nature, is the rule and not the exception, structures ecosystems, and greatly increases diversity. This is a corollary of the ABH because it is consistent with the fact that all species make their ecosystem better for life and other species, resulting in an increase in biodiversity.

References

Thompson, J. N. (2005). *The Geographic Mosaic of Coevolution.* Univ. of Chicago Press. ISBN: 9780226797625

Doebeli, M. & Knowlton, N. (July 1998). The evolution of interspecific mutualisms. *PNAS USA* 95 (15): 8676–80.

Lewis, D. H. (1973). Concepts in fungal nutrition and the origin of biotrophy. *Biol. Rev.* 48: 261-78.

Trappe, J. M. (1987). Phylogenetic and ecologic aspects of mycotrophy in the angiosperms from an evolutionary standpoint. *Ecophysiology of VA Mycorrhizal Plants*, G.R. Safir (Eds.). CRC Press, FL.

Newman, E. I. & Reddell, P. (1987). The distribution of mycorrhizas among families of vascular plants. *New Phytologist* 106: 745-51.

Pirozynski, K. A. & Malloch, D. W. (March 1975). The origin of land plants: a matter of mycotrophism. *Biosystems* 6 (3): 153-64.

Pirozynski, K. A. (1981). Interactions between fungi and plants through the ages. *Canadian Journ. of Botany* 59 (10): 1824-7.

Stubblefield, S. R., Thomas, N., Taylor, T. N., & Trappe, J. M. (Dec., 1987). Vesicular-Arbuscular mycorrhizae from the Triassic of Antarctica. *Amer. Journ. of Botany.* 74 (12): 1904-11. Stable URL: http://www.jstor.org/stable/2443974

Crepet, W. L. (1983). Chapter 3, The role of insect pollination in the evolution of the angiosperms, in Real, L., editor, *Pollination Biology*, Academic Press, Inc. (Harcourt, Brace Jovanovic, San Diego, CA, Publishers).

Seaborg, D. (2022). *How Life Increases Biodiversity: An Autocatalytic Hypothesis.* CRC Press; Taylor & Francis Group. Boca Raton, FL; London, U.K.; New York, NY.

William, M. & Koonin, E. V. (2006). Introns and the origin of nucleus–cytosol compartmentalization. *Nature* 440 (7080): 41-5.

Overmann, J., & Schubert, K. (2002). Phototrophic consortia: model systems for symbiotic interrelations between prokaryotes. *Arch. Microbiol.* 177201-208.

Margulis, L. & Sagan, D. (1986). Origins of Sex. *Three Billion Years of Genetic Recombination.* Yale University Press, New Haven, CT. pp. 69–71, 87. ISBN 0 300 03340 0

Martin, W. F. & Müller, M. (2007). *Origin of Mitochondria and Hydrogenosomes.* Springer Verlag, Heidelberg, Germany.

Cornejo-Castillo, F. M., Inomura, K., Zehr, J. P., & Follows, M. J. (11 March 2024). Metabolic trade-offs constrain the cell size ratio in a nitrogen-fixing symbiosis. *Cell* 187 (7): 1762-1768.E9. https://doi.org/10.1016/j.cell.2024.02.016

Hastings, J.W. (1978). Bacterial bioluminescence: An overview. *Methods in Enzymology.* 57: 125-35.

Haddock, S. H. D., Moline, M. A., & Case, J. F. (2010). Bioluminescence in the sea. *Ann. Rev. of Marine Sci.* 2: 443–93. doi: 10.1146/annurev-marine-120308-081028

Spribille, T., et al. (29 July 2016). Basidiomycete yeasts in the cortex of ascomycete macrolichens. *Science* 353 (6298): 488-92. DOI: 10.1126/science.aaf8287

Grube, M., Cardinale, M., Vieira de Castro, Jr, J., Müller, H., Berg, G. (25 June 2009). Species-specific structural and functional diversity of bacterial communities in lichen symbioses. *The ISME Journal* 3 (9): 1105–15. https://doi.org/10.1038/ismej.2009.63

Davies, A. B., et al. (May 2015). Seasonal activity patterns of African savanna termites vary across a rainfall gradient. *Insectes Sociaux* 62 (2): 157-65.

Faeth, S. H. (2002). Are endophytic fungi defensive plant mutualists? *Oikos* 98: 25-36.

Speight, M. R., et al. (1999). *Ecology of Insects.* Blackwell Science, Oxford, UK; p. 156. ISBN 0-86542-745-3.

Leal, I. R. & Oliveira, P. S. (1998). Interaction between fungus-growing ants (Attini), fruits, and seeds in cerrado vegetation in southeast Brazil. *Biotropica* 30: 170–8.

Zhang, M. M., Poulsen, M., & Currie, C. R. (2007). Symbiont recognition of mutualistic bacteria by Acromyrmex leaf-cutting ants.

The ISME Journal 1 (4): 313–30. doi: 10.1038/ismej.2007.41

Currie, C. R., Scott, J. A., Summerbell, R. C., & Malloch, D. (1999). Fungus-growing ants use antibiotic-producing bacteria to control garden parasites. *Nature* 398: 701-4. doi: 10.1038/19519

Vasconcelos, H. L. & Fowler, H. G. (1990). Foraging and fungal substrate selection by leaf-cutting ants. In: Vander Meer, R. K., JaVe, K., & Cedeno, A. (eds.) *Applied Myrmecology—a World Perspective*. Westview Press, Boulder, CO: 411–9.

Wirth, R., Beyschlag, W., Herz, H., Ryel, R. J., & Hölldobler, B. (2003). Herbivory of leaf-cutter ants: a case study of Atta columbica in the tropical rainforest of Panama. *Ecol. Stud.* 164: 1–233.

Silva, P. S. D., Bieber, A. G. D., Leal, I. R., Wirth, R., & Tabarelli, M. (2009). Decreasing abundance of leaf-cutting ants across a chronosequence of advancing Atlantic forest regeneration. *Journ. Trop. Ecol.* 25: 223–27.

Cherrett, J. M. (1989). Leaf-cutting ants., in: Lieth, H. & Werger, M. J. A. (eds.) *Ecosystems of the World*. Elsevier, New York: 473–86.

Harrison, M. J. (2005). Signaling in the arbuscular mycorrhizal symbiosis. Annual Rev. *Microbio*l. 59: 19–42. doi: 10.1146/annurev.micro.58.030603.123749. PMID 16153162

Smith, S. E. & Read, D. (2008). *Mycorrhizal Symbiosis*. (Third Edition). Elsevier Ltd., Phila., PA. ISBN: 978-0-12-370526-6.

Wang, B., & Qiu, Y.L. (2006). Phylogenetic distribution and evolution of mycorrhizas in land plants. *Mycorrhiza* 16 (5): 299–363. doi: 10.1007/s00572-005-0033-6. PMID 16845554

Simard, S.W., Beiler, K.J., Bingham, M.A., Deslippe, J.R., Philip, L.J., & Teste, F.P. (2012). Mycorrhizal networks: Mechanisms, ecology and modeling. *Fungal Biol. Rev.* 26: 39-60.

Andrade, G., Linderman, R. G. & Bethlenfalvay, G.J. (1998-1). Bacterial associations with the mycorrhizosphere and hyphosphere of the arbuscular mycorrhizal fungus Glomus mosseae. *Plant Soil* 202: 79-87.

Andrade, G., Mihara, K. L.,Linderman, R. G., & Bethlenfalvay, G. J.. (1998-2). Soil aggregation status and rhizobacteria in the mycorrhizosphere. *Plant Soil*. 202: 89-96.

Founoune, H., Duponnois, R., Meyer, J. M., Thioulouse, J., Masse, D., Chotte, J. L., & Neyra, M. (2002). Interactions between ectomycorrhizal symbiosis and fluorescent pseudomonads on

Acacia holosericea: isolation of mycorrhiza helper bacteria (MHB) from a Soudano-Sahelian soil. *FEMS Microbiol. Ecol.* 41: 37-46.

Garbaye, J. (1991). Biological interactions in the mycorrhizosphere. *Experientia* 47: 370-5.

Garbaye, J. (1994). Helper bacteria: a new dimension to the mycorrhizal symbiosis. *New Phytol.* 128: 197-210.

Frey-Klett, P., Garbaye, J., & Tarkka, M. (30 Aug. 2007). The mycorrhiza helper bacteria revisited. *New Phytologist.* https://doi.org/10.1111/j.1469-8137.2007.02191.x

Read, D. J., Duckett, J. G., Francis, R., Ligrone, R., & Russell, A. (29 June 2000). Symbiotic fungal associations in 'lower' land plants. Philosophical Transactions of the Royal Society B. *Biological Sciences* 355 (1398). https://doi.org/10.1098/rstb.2000.0617

Remy, W., Taylor, T. N., Hass, H., & Kerp, H. (6 Dec. 1994). Four hundred-million-year-old vesicular arbuscular mycorrhizae. *PNAS USA* 91 (25) 11841-43. https://doi.org/10.1073/pnas.91.25.11841

Simon, L., Bousquet, J., Lévesque, R. et al. (1993). Origin and diversification of endomycorrhizal fungi and coincidence with vascular land plants. *Nature* 363: 67–69. https://doi.org/10.1038/363067a0

Smith, M. R. (April, 2016). Cord-forming Palaeozoic fungi in terrestrial assemblages. *Botanical Journ. of the Linnean Soc.* 180 (4): 452–460. https://doi.org/10.1111/boj.12389

Crepet, W. L. (1984). Advanced (constant) insect pollination mechanisms: pattern of evolution and implications vis-a-vis angiosperm diversity. *Annals of the Missouri Botanical Garden.* Historical Perspectives of Angiosperm Evol. 71 (2): 607–30.

Regal, P. J. (1977). Ecology and evolution of flowering plant dominance. *Science* 196: 622–9.

Stebbins, G. L. (1981). Why are there so many species of flowering plants? *BioScience* 31: 573–7.

McPeek, M. A. (1996). Linking local species interactions to rates of speciation in communities. *Ecol.* 77: 1355–66.

Koh, L. P., et al. (10 Sept. 2004). Species coextinctions and the biodiversity crisis. *Science* 305 (5690): 1632-4. doi: 10.1126/science.1101101

Thorne, R. F. (2002). How many species of seed plants are there? *Taxon* 51 (3): 511–22. doi: 10.2307/1554864. JSTOR 1554864

Scotland, R. W. & Wortley, A. H. (2003). How many species of seed plants are there? *Taxon* 52 (1): 101–4. doi: 10.2307/3647306. JSTOR 3647306

Govaerts, R. (2003). How many species of seed plants are there?--a response. *Taxon* 52 (3): 583–4.

Crane, P. R., Friis, E. M., & Pedersen, K. R. (1995). The origin and early diversification of angiosperms. *Nature* 374: 27–33.

Dafni, A., Kevan, P. G., & Husband, B. C., eds. (2005). *Practical Pollination Biology*, Enviroquest, Ltd., Cambridge, Ontario, Canada.

Baker, H. G. & & Hurd, P. D. (1968). *Intrafloral Ecology. Annu. Rev. Entomol.* 13: 385-414.

Ollerton, J., Winfree, R., & Tarrant, S. (March, 2011). How many flowering plants are pollinated by animals? *Oikos* 120 (3): 321–6. doi: 10.1111/j.1600-0706.2010.18644.x

Landry, C. L. (2010). Mighty mutualisms: The nature of plant-pollinator interactions. *Nature Education Knowledge* 3 (10): 37.

Hammond, P. M. (1992). Species inventory; in *Global Biodiversity, Status of the Earth's Living Resources*, pp. 17-39. Groombridge, B., ed. Chapman and Hall, London, UK. ISBN 0412472406.

Capinera, J. L. (2008). Butterflies and Moths. *Encyclopedia of Entomol.* 4 (2nd ed.); Springer, New York. N. Y., pp. 626–72. ISBN 9781402062421.

Mallet, J. (12 June 2007). Taxonomy of Lepidoptera: the scale of the problem. *The Lepidoptera Taxome Project*. University College, London.

Van der Pijl, L. (1982). *Principles of Dispersal in Higher Plants*, third revised and expanded edition; Springer-Verlag, Berlin, Heidelberg; Germany; New York, N. Y.

Stoner, K. E. & Henry, M. (2009). Seed dispersal and frugivory in tropical ecosystems. *Tropical Biol. And Conservation Management V. Encyclopedia of Life Support Systems (EOLSS)*. Internat. Comm. on Tropical Biol. and Nat. Resources.

Bello, C. & Galetti, M., Pizo, M. A., Fernando, L., Magnago, S., Rocha, M. F., et al. (18 Dec. 2015). Defaunation affects carbon storage in tropical forests. *Science Adv.* 1:e1501105; pp. 1-10.

Spaulding, M.D., Ravilious, C, & Green, E.P. (2001). *UNEP-WCMC World Atlas of Coral Reefs*. Prepared at the UNEP World Conservation Monitoring Centre. University of California Press. Berkeley, USA.

Sherman, C.D.H. (2009). *The Importance of fine-scale environ-*

mental heterogeneity in determining levels of genotypic diversity and local adaption. University of Wollongong Ph.D. Thesis.

Marshall, P. & Schuttenberg, H. (2006). *A Reef Manager's Guide to Coral Bleaching.* Great Barrier Reef Marine Park Authority. Townsville, Australia. ISBN 1-876945-40-0.

Spalding, M.D. & Grenfell, A. M. (1997). New estimates of global and regional coral reef areas. *Coral Reefs* 16 (4): 225–30. doi: 10.1007/s003380050078

Kaplan, M. (2009). How the sponge stays slim. *Nature.* doi: 10.1038/news.2009.1088

McKeon, C. S. & Jenna M. Moore, J. M. (30 Sept. 2014). Species and size diversity in protective services offered by coral guard-crabs. *PeerJ* 2:e574. https://doi.org/10.7717/peerj.574. PubMed 25289176

Mallet, J. & Joron, M. (1999). Evolution of diversity in warning color and mimicry: polymorphisms, shifting balance, and speciation. *Annu. Rev. Ecol. Syst.* 30: 201-33.

Futuyma, D. J. (2005). *Evolution.* Sinauer Assoc., Inc., Sunderland, MA. p.286, fig.12.19 and p. 445, Fig.18.24.i

Lago, J., et al. (2015). Tetrodotoxin, an extremely potent marine neurotoxin: Distribution, toxicity, origin and therapeutic uses. *Marine Drugs* 13 (10): 6384–406. doi: 10.3390/md13106384. PMC 4626696. PMID 26492253

Heinrich, B., Stahler, D., & D. Smith, D. (2002). Common ravens preferentially associate with grey wolves as a foraging strategy in winter. *Animal Behaviour* 64: 283–90. doi: 10.1006/anbe.2002.3047

Vucetich, J. A., Peterson, R. O., & Waite, T. A. (2003). Raven scavenging favours group foraging in wolves. *Animal Behaviour* 67: 1117–26. doi: 10.1016/j.anbe hav.2003.06.018

Nardi, J.B., Mackie, R. I., & Dawson, J. O. (2002). Could microbial symbionts of arthropod guts contribute significantly to nitrogen fixation in terrestrial ecosystems? *Journ. of Insect Physiology* 48: 751-63.

Raine, N. E., Willmer, P., & Stone, G. N. (Nov. 2002). Spatial structuring and floral avoidance behavior prevent ant-pollinator conflict in a Mexican ant-acacia. *Ecol.* 83 (11): 3086–96. doi: 10.1890/0012-9658(2002)083[3086:SSAFAB]2.0.CO;2

González-Teuber, M., et al. (April 2014). Mutualistic ants as an indirect defense against leaf pathogens. *New Phytologist* 202 (2): 640-50. doi: 10.1111/nph.12664

Dorenbosch, M., et al. (2005). Indo-Pacific seagrass beds and mangroves contribute to fish density and diversity on adjacent coral reefs. *Mar. Ecol. Prog. Ser.* 302: 63-76.

CHAPTER 6. COMMENSALISM IS UBIQUITOUS, AND MAINTAINS AND INCREASES BIODIVERSTY

Introduction

Commensalism is a relationship between two species in which one species receives benefits and the other is unaffected. I will refer to the species that helps as the benefactor, and the species that is helped as the beneficiary. Commensalism is a major mechanism by which the ABH/Pachamama Hypothesis is carried out, as it maintains and even can greatly increase species diversity. One species either merely benefits from or cannot exist without the unaffected species. Countless species are able to persist only because of benefactor species. The more common commensalism is, the more important it is to biodiversity and the more the ABH is supported. It is not known precisely how common it is, although the majority view that it is very common. It is almost certainly the most common interspecific interaction when one takes into account the fact that all complex organisms are ecosystems harboring a huge number of micro-organisms, called the microbiome, that they benefit by providing them with a habitat. Many of these microbes are symbiotic, helping the host organism, but most might be commensal, having no effect of their host. Commensalism maintains and promotes biodiversity because one species helps another without being harmed. Commensalisms vary in strength and duration from intimate, long-lived relationships to brief, weak interactions, sometimes through intermediaries.

This chapter lists some examples of commensalism, demonstrating how common it is and how it maintains and increases biodiversity.

Beneficiary Commensal Species that Live in
the Dwellings of, or in or on, Other Species

The abandoned nest holes of many woodpecker species are used by several other bird species. In the southwest U. S., the elf owl uses the hole in a saguaro cactus made by the Gila woodpecker for nesting after the woodpecker is done nesting in it.

The lace monitor, a large Australian lizard, lays its eggs in termite nests. Since the temperature in the nests are well regulated by the termites to be neither too hot nor too cold, the eggs do well. When the babies hatch, they are too small to burrow out of the nest. The female returns at this time and digs them out.

The pygmy blue-tongued skink of Australia lives in trap door spider burrows and lays its eggs in them. The young stay there a while after hatching before leaving.

In Argentina, lizards lay eggs inside the nests of the caimans, which are relatives of alligators. This is safe because the mother caiman does not feed during incubation of her eggs. The lizard's eggs get protection and their own nest.

Many beneficiary species use other species, or parts of them, as habitat. Eyelash mites live at the base of human eyelashes, normally causing no harm. Hermit crabs use the shells of dead sea snails as homes. The snail is dead and so unaffected. They are also symbiotic with sea anemones that grow on the snail's shell. The anemone gets transportation and provides protection from its stinging tentacles to the hermit crab. And the anemone is a beneficiary commensal of the sea snail, being in the same situation as the hermit crab and snail.

Some one-celled organisms use silicon or calcium shells of phytoplankton (such as those of coccolithophors) to construct their own armor.

161

Epiphytes are small plants that grow on large plants such as trees. They benefit mainly by gaining height and access to sunlight for photosynthesis. They also gain access to a greater number of animal pollinators higher up on the tree, often at the top of the forest. They can more easily disperse their seeds via wind. Host plants also benefit epiphytic plants by providing them with moisture. Epiphytes do obtain moisture from rain and the air, but they also get it from the surface of the host plant. Finally, they are out of reach of ground-dwelling herbivores. The benefactor tree is generally unaffected, so it is usually commensalism, but sometimes the epiphytes become so plentiful that their weight hurts the host tree, in which case the relationship is parasitism.

Some plants groups are specialized as epiphytes or have many species that are so specialized; these include the bromeliads and orchids. There are over 2,000 species of epiphytes that are in the bromeliad family. Seventy per cent of orchids are epiphytes, and orchids are the most diverse family of flowering plants, with over 18,000 species, representing about 8% of all flowering plants on Earth. Furthermore, it is estimated that there are 10,000 to 12,000 orchid species that have yet to be described. This yields about 12,600 described species of epiphytic orchid, and an estimate of about 19,600 to 21,000 total epiphytic orchid species, when undescribed species are added! There are epiphytes in 83 plant families. Striking evidence for the prevalence of commensalism comes from the estimate that 24 to 25% of all land plants are epiphytes (Almeda, 2012). There are a number of cacti that are epiphytes, although they represent a minority of this family.

Some epiphytes act as benefactor species. They create niches that are exploited by a wide range of species. One of the best examples of a tiny ecosystem in an epiphyte is the tank bromeliad of South America, whose stiff, upturned leaves can hold more than two gallons (eight liters) of water. These reservoirs of water not only provide a drinking supply for many tree-top animals, but also create an entire habitat which species use for living and breeding. A multitude of insect larvae exist in these pools and are fed upon by other animals. And these pools in water of the tank bromeliad serve as a nursery for poison-arrow frog tadpoles.

Commensalism Is Ubiquitous, and
Maintains and Increases Biodiversity

Pearlfish are small fish that in most species live as adults inside invertebrates, mainly clams, sea stars, or sea squirts, as commensals, not harming their hosts. Some species are parasites, living in sea cucumbers, bivalves, and starfish, eating their insides. The ones that live in sea cucumbers live in the anus and eat the gonads.

On the rocky shores of western North America, the California mussel (*Mytilus californianus*) forms extensive mussel beds. Hundreds of invertebrate species, including barnacles, polychaetes (marine worms), and snails live on and in the interstices of the mussels, which function analogously to trees in a rainforest. Mussels provide a three-dimensional habitat in which to live, but they also provide a wet microclimate with food for commensal beneficiaries.

Batesian Mimicry

Batesian mimicry is a relationship where one species, the mimic, resembles an unpalatable species, the model. The model may be poisonous and distasteful to eat, or have a venomous bite or sting, for example. Model species have warning color patterns, sounds, or smells that they may reinforce with behaviors. The mimic is harmless, and benefits from the model by imitating its warning colors, sounds, smell, and/or behaviors. The model is mostly unaffected, although it acquires a small cost because the predator has a harder time learning that the model is unpalatable, since it encounters the palatable mimic once in a while. Sometimes the predator does not need to learn to avoid the unpalatable species, but instinctively avoids it. The percentage of all animal species that are Batesian mimics is not known. It is thought to be more common in the tropics.

There are species of mimics that vary considerably, in which there are several color patterns in one species, each pattern resembling a different model species. This mimicry occurs especially in beetles and butterflies. In fact, mimicry is especially common in butterflies and moths.

There are many types of insect mimics, and mimics may resemble models that are wholly unrelated. There are flies and moths that mimic bees, moths that resemble hummingbirds, ants that look like tufts of thistle fluff, chrysalids of butterflies that look like twigs and leaves, and much more.

In snakes, the false cobra is a mildly venomous but harmless snake which mimics the characteristic hood (large, flattened neck) of the highly venomous Indian cobra's threat display. There re a number of species of harmless snake that resemble various species of the highly venomous coral snake.

The mimic octopus (*Thaumoctopus mimicus*), a Batesian mimic of the Indo-Pacific is likely the best mimic of all, capable of changing color and texture to blend cryptically with the environment and to mimic up to 15 species, many poisonous, more than any other animal can. It can change shape and take on the appropriate behavior to reinforce the mimicry. Some of the common animals it mimics are lion fish, where its legs resemble spines; sea snake, by waving two arms in opposite directions, with the yellow and black markings of the snake; flatfish, by flattening out its body while moving forward along the sea floor; sea squirts; sponges; jellyfish, imitating their motions; and even algae-covered rock coral. It mimics the animal appropriate to the predator it encounters.

Phoresy

Phoresy is commensalism in which the larger host species transports the smaller species, called the guest, without affecting the host. The host is not significantly affected in any of the following examples. Phoresy mainly occurs in arthropods. The beetle *Nymphister kronaueri* uses its mandibles to attach between the thorax and abdomen of the army ant *Eciton mexicanum* to obtain transport in the rainforests of Costa Rica (von Beeren and Tishechkin, 2017). Both the beetle and ant have no common name. The beetle mimics the ant's abdomen. Other hitchhiker species ride on army ants' backs, follow in their wake on foot, or stow themselves on top of prey the ants carry from nest to nest.

Pink whiprays, small stingrays of the tropical Indo-Pacific from southern Africa to Polynesia, hitch rides on larger species of stingrays. Remoras are fish of several species that have dorsal fins on top of them modified as suckers to attach to and get transport from sharks and other large fish. They eat food scraps from the prey of the sharks and likely get protection from them.

Various mite species get from flower to flower on the beaks of pollinating hummingbirds, or get transport on dragonflies, beetles, flies, and bees.

Other Examples of Commensalism

About 18 of the 200 species of antbirds specialize in following columns of army ants to eat the small invertebrates flushed up by the ants.

Pirt and Lee (1983) found that methane-generating Methanobacteria of the genera *Methanobacterium*, *Methanococcus*, and *Methanosarcina*, found in certain ponds and swamps, are commensal with and dependent on mobile bacteria that consume oxygen and produce carbon dioxide. The Methanobacteria cannot live in the presence of oxygen. So the oxygen-consuming bacteria help the Methanobacteria by removing oxygen from their environment and providing them with the energy source, carbon dioxide. The oxygen-consuming bacteria do not benefit from the Methanobacteria. This happens in wetlands of various depths and sediment compositions.

Western honey bees collect chemicals from mushrooms that protect them against viral infections (Stamets et al., 2018).

Some insects acquire compounds from their host plants and use them as perfumes or precursors to them (Reddy and Guerrero, 2004, and references therein).

The black giant cicada is an insect that lives in high densities in urban parks in central Japan in summer. They pierce holes in keyaki trees with their mouthparts, then feed on the exuding sap. Three ant species, two wasp species, and two species of flower chafer beetles, feed on the sap after the cicada finishes feeding (Yamazaki, 2007).

Butterflies in the western Amazon land on and drink the liquid in the eyes of yellow-spotted river turtles to obtain sodium, perhaps other minerals, and possibly amino acids.

Keystone Species

A keystone species is one that has a disproportionately large positive effect on other species relative to its abundance and compared to the effect an average species has on other species. If it is removed, the ecosystem loses many species or even collapses. It is the commensal benefactor of a great many species, as well as being in symbiotic relationships with several species. Keystone species play a key role in determining their community's structure, and greatly increase and maintain their community's diversity. Many are keystone species because they provide habitat for immense numbers of species. The concept has been rightly criticized for oversimplifying complex ecological systems (Mills et al., 1993). Still, the concept has some validity.

Nitrogen-fixing bacteria convert nitrogen in the atmosphere, which plants cannot use, to a form that plants can employ. They provide useable nitrogen for multitudes of organisms. Nitrogen is a very important essential element for life, used in DNA, RNA, and protein. Although symbiotic with their plant allies, which provide them with a home in their roots, these bacteria commensally benefit tens of thousands of species indirectly by helping plants. Cyanobacteria fix nitrogen in the sea, helping countless species.

Thousands of coral reef species benefit from the three-dimensional habitat provided by several keystone coral species. This includes hundreds of fish and thousands of invertebrate species. There are several commensal invertebrate species that live in the coral, either boring into it or inhabiting pre-existing crevices. Boring animals include sponges and bivalve mollusks. Animals that live in crevices include crabs and worms.

Sea birds benefit from islands built by dead coral, including for nesting. There are 17 species comprising a total of almost three million seabirds on Midway Atoll, which is made of coral, including 1.5 million Laysan albatross, which is two-thirds of the global population. One-third of the worldwide population of the black-footed albatross also occur on Midway Atoll.

The jaguar is a keystone species in Central and South America because it has an extremely varied diet, helping to balance the jungle ecosystem with 87 different prey species. This not only benefits the prey populations that are regulated, but many mammals, birds, invertebrates, and plants that benefit because their predators and herbivores are regulated.

The cassowary, a large, flightless bird of the tropical forests of New Guinea, nearby islands, and north-eastern Australia, of which there are three living species, eats fruit. Each cassowary species is a keystone species. They disperse the seeds of up to 238 rainforest plant and tree species. Without cassowaries, the rainforests in which they live would not be able to survive. It is estimated that up to 100 plant species depend exclusively on cassowaries to disperse their seeds. Cassowaries are the only species that can digest some species of fruit. This includes the cassowary plum (*Cerbera floribunda*). Some fruits will not grow unless they have been through a cassowary's digestive system. This keystone species indirectly aids all the species that benefit from the many tree species whose seeds it disperses.

In North America, the grizzly bear is a keystone species because it transfers nutrients from the sea to the forest. Salmon, rich in nitrogen, sulfur, carbon, and phosphorus, swim up rivers, sometimes for hundreds of miles. The bears capture the salmon to eat them, carrying them onto dry land, defecating and dispersing partially-eaten fish. It has been estimated that the bears leave up to half of the salmon they harvest on the forest floor. This benefits a huge number of species of land plants (including trees), which benefit great numbers of animals. Salmon themselves also die of natural causes in tremendous numbers, aiding tree growth by providing nutrients to aquatic ecosystems. And some of the nutrients end up on land.

Many tree species, including figs, oaks, and rainforest species, are keystone species, providing habitat and/or food for a multitude of bacteria, fungi, plants, invertebrates, birds, and mammals, and some amphibians and reptiles. The beneficiary species live in or under the trees, or in the soil under the trees, and use the trees' shade, leaves, branches, trunk, and/

or roots, depending on the beneficiary. Trees provide a three-dimensional habitat by adding height. The canopy of the rainforest is the most diverse terrestrial ecosystem, mainly because of insects inhabiting it, but it also houses all groups mentioned in the first sentence of this paragraph. Trees provide shade and shelter for ground-dwellers, and food for herbivores. Predators that eat these herbivores benefit indirectly from trees. Dead trees provide food and habitat for many small vertebrate and invertebrate species, some of which live under rotting logs. Insects likely have the most commensal beneficiary species of any animal class on trees. The number of insect species in an acre of tropical rainforest canopy could approach tens of thousands or more, and sampling of a tropical rainforest canopy found such high insect diversity that it was proposed the total estimated number of animal species on Earth be increased from 5 to 10 million to 30 to 60 million (Erwin, 1983). Trees also provide shade that protects certain plant species from too much sun. They also lower the temperature of rivers by shading them, allowing certain fish, such as salmon and trout, and invertebrates to live in the rivers.

There are an estimated 16,000 tree species in the Amazon rainforest. There can be a hundred species of tree per acre in rainforests. Each of these tree species provides habitat or other benefits to many vertebrates, hundreds of invertebrate, and tens of thousands of fungal and microbial species. Each of these species are found on a limited number of tree species, so the diversity of species benefiting from an acre of rainforest trees is astronomical. Animal species commensal or symbiotic with rainforest trees number at least in the hundreds of thousands.

Every species of tree is a mini-ecosystem, with amazingly far-reaching impacts on its community and ecosystem, greatly increasing species diversity in its community. Many of the beneficiary species aid trees and are thus symbiotic with trees, but many are commensal. The number of species that benefit from any given tree species is astronomical.

Tress also remove carbon dioxide from the air; produce oxygen; provide habitat for epiphytes; hold the soil in place; create rain; and provide other services. Many plant and animal species in all types of forests

168

benefit from the shade of trees. Trees support small ecosystems. All trees (in fact, all plants) harbor very diverse ecosystems of microbes called the plant microbiome. The number of species that benefit from any given tree species is astronomical.

It was land plants, which first appear in the fossil record about 400 mya, that made it possible for invertebrates and amphibians to evolve and colonize the land. Plants provided food for invertebrates. They built the soil. Their roots held soil in place and allowed the formation of stable banks for streams and ponds (Gibling and Davies, 2012)—habitat for invertebrates and amphibian larvae. Invertebrates were able to colonize land only after plants were there, and did so soon after land plants appeared. They were a necessary food source for amphibians. So amphibians needed both land plants and invertebrates to come onto the land.

Many species of earthworms help the soil. In many soils, worms pull fallen leaves and manure lying on the ground down below the surface, for food or to plug their burrows. In their burrows, underground, the worms nibble the leaf and partially digest it, saturate it with intestinal secretions, and mix it into the soil. Thus, the worms convert organic matter such as leaves and manure on the surface into rich humus in the soil, greatly improving its fertility. Humus is the organic component of soil, important for supporting soil life.

Earthworms also eat soil particles. These are ground in the gizzard by tiny pieces of grit into a fine paste which the intestine then digests. The material is excreted in what are called casts, on the surface or deeper in the soil. The casts contain minerals and nutrients that are accessible to plants, but were not obtainable before the worms put them through this process. Earthworm casts are five times richer in available nitrogen, seven times richer in available phosphates, and eleven times richer in available potash than the surrounding upper six inches (150 millimeters) of soil. They can contain 40% more humus than the top nine inches (23 centimeters) of soil near where they live. In conditions with abundant humus, one worm can produce ten pounds (4.5 kilograms) of casts per year. The worms mix the soil, and their actions help plants take up minerals and nutrients.

The worms create a great deal of channels by burrowing. This allows aeration and drainage of the soil, essential to its health and useful to life. They even secrete mucous that lubricates the soil, making movement easier. The worms act as pumps, pumping air in and out of the soil, more rapidly at night. So the worms not only create passages for air and water to move through soil, but are also active pistons to move the air.

Their impact is great, for their numbers are huge. Even poor soil may support 250,000 earthworms/acre (62/square meter), while rich, fertile farmland may have up to 1,750,000/acre (432/square meter). The weight of earthworms beneath the soil of a farm is likely greater than that of the livestock on its surface.

Therefore, earthworms help all species that benefit from soil. This includes myriad species of plants, especially trees, the mycorrhizal fungi and nitrogen-fixing bacteria associated with the roots of trees and some plants, thousands of species of soil invertebrates, and burrowers like moles and gophers. It also includes huge numbers of species that benefit indirectly from the soil, such as various species of birds and squirrels that benefit in one way or another from trees or plants, and the predators of these birds and squirrels, such as birds of prey, coyotes, foxes, pine martins, raccoons, snakes, and so on. Other indirect earthworm beneficiaries include insects and other invertebrates above ground that feed on the plants, and their predators. I estimate that this amounts to hundreds to thousands of species in temperate regions, and tens to hundreds of thousands of species in the tropics.

There are several other invertebrate species that benefit the soil and thus several plant and animal species.

Oysters, mussels, scallops, and some other mollusks modify their environment, making it more favorable to biodiversity. They filter out pollutants, including industrial waste, and algae, cleaning the water. A single oyster can filter and clean 50 gallons of water per day. Freshwater mussels clean rivers; a single one can filter over 50 liters of water per day. Mussels and oysters anchor sediment, preventing erosion. Freshwater mollusk shells stabilize river beds. Mussels and oysters protect coastal habitats from daily tides and storm surges. They provide a three-dimensional habitat for several fish and invertebrate species. They are food for sea stars, fish, some snails, and other species.

Mangroves protect coastal habitats by reducing erosion from tides and dampening the energy of storm waves. They prevent erosion and hold soil in place. They even accumulate and build soil in their tangled roots, creating habitat for themselves. They filter out toxins. They provide habitat for many fish and invertebrate species. They provide nurseries for the larvae of coral and some coral reef fish. They reduce atmospheric carbon dioxide by drawing it out of the air and burying the carbon in soil.

Prairie dogs, gophers, desert and gopher tortoises, American badgers, ground squirrels, mongooses, and other animals that dig burrows have many animals that use their burrows temporarily or permanently, including insects, scorpions, other arthropods, salamanders, frogs, toads, lizards, snakes, and rodents. The burrows also house fungi, which some of the insects eat. Gopher tortoise burrows are used by over 300 species in long-leaf pine forests of the US southeast. Besides using them to regulate their temperature, animals seek refuge from forest fires in them. The California tiger salamander (*Ambystoma californiense*) utilizes and is largely dependent on the burrows of the California ground squirrel (*Otospermophilus beecheyi*). The benefactor species in this paragraph till the soil, and their burrows aerate the soil. Echidnas, Australian egg-laying mammals, till and aerate the soil, helping a multitude of species. They spend about 12% of their day digging, mainly for invertebrates, but at times to escape predation. One echidna churns up 204 cubic meters (about 7,204 cubic feet) of soil in one year (Clemente et al., 2016).

Prairie dogs of several species in the western U. S. are important keystone species. They dig burrows for themselves that are employed as nesting sites for burrowing owls and birds called mountain plovers, and provide habitat for various species of snakes, lizards, salamanders, frogs, toads, mice, other rodents, and invertebrates, allowing them to escape the heat and cold. Burrowing owls that use the burrows flee predators in response to the prairie dogs warning whistles. Their tunnel systems help channel rainwater into the water table, preventing runoff and water loss, as well as erosion. Their tunnels also act as channels for air to go into the soil, increasing aeration. This is good for the soil and its ability to support life. Prairie dogs till and loosen the soil, reversing soil compaction, a deleterious effect on the soil that can be caused by grazing hooved mammals such as deer. The soil compaction is not a counter example to the ABH, because this effect is always ameliorated by burrowing animals that coexist with the hooved mammals, and these mammals improve the soil for life by tilling it. Prairie dogs trim the vegetation around their colonies, perhaps to better see their predators. They also eat back the vegetation, trimming it short as they graze. The grasses and other vegetation grow back with much higher nutritional content after prairie dog trimming. This greatly helps grazing hoofed mammals, such as American bison, pronghorn, and mule deer. These three species have shown a proclivity for grazing areas trimmed by prairie dogs. This is interesting, because these species are competitors with the prairie dog, eating many of the same plants. Yet, they benefit more from the prairie dog than they are harmed by it. They are both beneficiary commensals and competitors of prairie dogs.

The burrows of all animals discussed above regulate atmospheric temperature because their burrows aerate the soil, and the oxygen they bring to the soil inhibits the growth of methanogens, which cannot grow in the presence of oxygen and produce the potent greenhouse gas, methane. The oxygen in the burrows also combines with and depletes any methane the methanogens do produce there.

Bioturbation is the reworking of soils and sediments, including aquatic ones, by animals or plants. These include burrowing, and ingestion and defecation of sediment. Generally, organisms that do this are keystone species, since it has profound positive effects on the environment and is a primary driver of biodiversity (Wilkinson et al., 2009). Bioturbers improve habitat for themselves and myriad other species. Bioturbation is a major component in the cycling of elements, including magnesium, nitrogen, calcium, strontium, and molybdenum. It increases availability to organisms of the mineral nutrients calcium, phosphate, iron, and sulfur. The terrestrial environment has been altered greatly by the actions of organisms throughout Earth's history (Schwartzman, 1999).

Multicellular animals altered the aquatic sediments by processing them through their guts and burrowing in them (McIlroy and Logan, 1999). This introduced water into the sediments, increasing their suitability as a habitat. Deeper burrowing became easier. Bioturbation aerates the sea floor sediments, enabling animals and bacteria that need oxygen to extend their habitat to deeper levels.

Bioturbation in the sea is done by walruses, worms, ghost shrimp, mud shrimp, and several other animals, and in fresh water systems by salmon, midge larvae, and many other fauna. Bioturbation in aquatic systems increases nutrients in aquatic sediment and overlying water; mixes water and nutrients in sediments, enhancing habitat; provides burrows; and improves sediment texture. The transport of nutrients like dissolved oxygen, and enhancement of organic matter decomposition and sediment structure, on land and in water, by bioturbators aids the survival and colonization by other animal and microbial communities (Braeckman et al., 2010).

Pillay (2010) pointed out that bioturbators potentially play a big role in evolution. It has likely profoundly influenced evolution ever since it started. It has been important in nutrient cycling, and bioturbating animals likely affected the cycling of sulfur in the early oceans in a way helpful to life. Stromatolites are microbial mats consisting of colonial microbes, mainly photosynthetic cyanobacteria. They grow to a meter or more in height. They are among the oldest life on Earth, and dominated the earliest life. They are less common today. Microbial mats--in particular, stromatolites--were the main biological structures in the sea before the Cambrian explosion, providing three-dimensional habitat, food, and oxygen for other life, but the habitat they provided was limited to the sea floor, and animals did not have access to the habitat below the sea floor. Prokaryotes on the sea floor caused organic matter to accumulate in the sediments. The bacteria and this organic matter were food sources for bottom-feeding animals. Predators were a selective force favoring hard skeletons and parts, including bristles, spines, and shells, in their prey.

This allowed burrowing, which the prey did both for protection and to feed on the organic matter in the sediments. Predators then evolved the ability to burrow into the sediments to seek their prey. Thus, bioturbation came about, and burrowing animals disturbed the microbial mat system and created a mixed sediment layer with greater biological and chemical diversity, which is thought to have led to the evolution and diversification of seafloor-dwelling species and the Cambrian explosion (Meysman et al., 2006; Pillay, 2010). This means the Cambrian explosion was made possible by bioturbation by burrowers creating habitat for more complex seafloor dwellers. This is significant because the Cambrian explosion was among the most important evolutionary diversifications in the history of life, creating to all the major animal phyla of today. In the Ediacaran Period, which consisted of simple animals without skeletons or shells and occurred from 635 mya to the Cambrian explosion, microbial mats provided food for many species, but hardened the seafloor, making it difficult for many animal species to utilize it. There was an expansion of burrowing animals from 550 to 540 mya (Grazhdanklin et al., 2012), possibly by primitive worms, just prior to and including the beginning of the Cambrian explosion. The burrowing seems to have opened up new living spaces for other animal species. Churning the mud would soften it and free it of the stiff microbial mat that covered much of the seafloor, making it easier to colonize. This would have had a planet-wide impact, creating habitat for many species, which could then diversify. Bioturbation heralded a new age in animal-sediment interaction and biogeochemical cycling, and made possible the Cambrian explosion.

Bioturbation is crucial in the deep sea because this habitat has little to no forces for sediment movement for nutrient cycling. On land, it is accomplished by burrowing mammals such as gophers, earthworms, burrowing insects, and several other species. Here, it aerates the soil, creates channels for water and air movement, and creates tunnels that animals can move through.

A lack of bioturbating organisms in coastal ecosystems results in clogging of the sediment with fine particles, a drastic reduction in sediment permeability, inability of oxygen to penetrate deeply into the sediment, and accumulation of reduced mineralized products in pore water. As a result, species diversity decreases greatly.

Elephants modify their ecosystem, making it more favorable to life. They dig water holes to get access to underground water when surface water is in short supply. A host of other animal species then use this water that was formerly unavailable to them. This includes all the major predators, herbivores, and scavengers that share the ecosystem with elephants. In Africa, for example, this is a tremendous number of species, including lions, cheetahs, leopards, various species of gazelle, giraffe, wildebeest, jackal, small mammals like mongoose, and diverse species of birds, reptiles, amphibians, aquatic insect larvae, flatworms, and other invertebrates. Elephants dig up earth to obtain salt and other minerals. Their tusks are used to churn the ground. The elephant eats dislodged pieces of soil, ingesting nutrients. The digging creates holes that are several feet deep, making salt and other minerals available to many other species. Animals such as rodents and reptiles use these holes as habitat. The footprints of African elephants in Kibale's tropical rainforest in Uganda fill with rain water and are used in the dry season by water beetles, mosquito larvae, dragonfly larvae, and other invertebrates, representing 61 distinct animal groups, including 27 families or orders (Remmers et al., 2017). Colonization of water-filled footprints is fast. They constitute important habitats with high diversity and variability, and they act as stepping stones for invertebrate dispersal. Platt et al. (2019) found water-filled tracks of the Asian elephant had frog and toad eggs and tadpoles. Their findings suggest that the tracks persist for over a year and function as small, still waterbodies that provide temporary, predator-free breeding habitat for frogs and toads during the dry season when alternate sites are unavailable. Trackways could also function as stepping stones that connect their populations. In India, Asian elephant footprints are used by mosquitoes to breed. All this suggests that the footprints of dinosaurs such as *Brontosaurus* made pools that were habitats for bacteria, archaea, protists, algae, fungi, insects and their larvae, amphibians and their larva, and perhaps other organisms in the time of dinosaurs. Elephants also eat and destroy trees, maintaining grasslands. This increases biodiversity, because it creates a mosaic of two habitats, forest and savanna, instead of only forest, even though forest is more diverse than savanna. Elephant defecation adds tremendous nutrient to the savanna, fertilizing it and keeping it healthy. This allows better growth of the grasses, and hence healthier herbivores and their predators. All three trophic levels are thus made more diverse.

Various species of copepods (small, numerous crustaceans found in nearly every freshwater and saltwater habitat); krill; shrimp, such as brine shrimp; and other small, oceanic groups cause great turbulence in the sea when they swim in great numbers, often by migrating from shallow to deeper waters and back. Brine shrimp can cause fluid velocities of about one to two centimeters per second. These various groups mix shallow waters with deeper, saltier waters. Without this mixing, waters of different densities and temperatures would remain separated in distinct layers, becoming stratified and stagnated. Parts of the sea would be very low in oxygen. The swarms mix ocean layers, and deliver nutrients from deep waters to sunlit shallow waters with phytoplankton, just as upwelling does. This increases phytoplankton, and hence all members of the food web above them. Phytoplankton growth helps sequester carbon. The swimmers also aid the exchange of gasses between sea and atmosphere; this helps life in a few ways, including making oxygen available to sea life. The animals are tiny, but extremely numerous. As one swims upward, it kicks the water downward. That parcel of water is then kicked down by another animal, and then another. This results in a downward flow of water that strengthens as the migration continues, and eventually extends as far as the entire migrating group, which can be as much as 100 meters (about 328 feet).

The beaver transforms parts of rivers to ponds or small lakes, creating habitat for great numbers of species. When a beaver family moves into new territory, they drop a large tree across a stream to begin a new dam, which creates a pond for their lodge.

This pooling of water leads to a cascade of ecological changes. The pond nourishes young willow, aspen, poplar, and cottonwood trees, the beaver's preferred trees for food and lodging. The pond aids the growth of grass and shrubs alongside of it, which improves habitat for songbirds, deer, and elk. Since dams raise underground water levels, they increase water supplies.

Commensalism Is Ubiquitous, and
Maintains and Increases Biodiversity

The dam increases habitat diversity by creating a pond and a wetland, resulting in three habitats--river, pond, and wetland--whereas before there was only the river. Thus, beavers increase biodiversity by increasing the number of habitat types, creating a mosaic of habitats. Species that can exist as a result of beaver ponds include several species of each of these: algae, plants that thrive in still water, insects with aquatic larvae, crayfish, other invertebrates, fish that prefer still or slow-moving water, frogs, toads, salamanders, some snakes, and water-loving birds. Indirect benefactors are all those species that are helped by the species the beaver benefits, such as snakes that eat frogs. This amounts to a huge number of beneficiary species that are commensal with the beaver, both directly and indirectly.

Beaver dams raise the water table alongside a stream, aiding the growth of trees and plants that stabilize the banks and prevent erosion. The trees provide shade that keeps the water temperature from getting too hot, allowing the survival of cool water species such as salmon.

Beaver dams have negative impacts on fish and wildlife, such as impeding salmon movement and decreasing their spawning habitat. But Kemp et al. (2012) found the number of benefits of dams to fish outweighed the number of costs, 184 to 119. The dams increase rearing and overwintering habitat, refuges, and invertebrate production. The latter increases species diversity in its own right, and also provides food for fish, frogs, and other taxa.

Beaver dams create nurseries for salmon, trout, and other fish. When beavers were made extinct in the Columbia River watershed in the early 1800's, salmon populations fell steeply in the following years. None of the other factors associated with the decline of salmon runs were extant at that time. Beaver dams increase salmon runs. The ponds they create are sufficiently deep for juvenile salmon to not be easily visible to predatory wading birds. They trap nutrients the juvenile salmon consume after losing their yolk sacs, including the bonanza nutrient pulse from the migration and then death of the adult salmon upstream. They create calm water with the result that the young salmon need not fight currents and can thus use their energy for growth. Thy keep the water clean. All fish of the salmon family are sensitive to water quality.

Beaver dams help frogs and toads because they provide areas with deep, warm, well-oxygenated water, and places for concealment for their larvae. Warm, oxygen-rich water enhances the development and growth of frog and toad larvae. Amphibian traps in beaver ponds in Alberta, Canada, captured 5.7 times more newly metamorphosed wood frogs, 29 times more western toads, and 24 times more boreal chorus frogs than on nearby free-flowing streams without beavers (Stevens, 2007).

Beaver dams help migrating songbirds, including ones in decline, by stimulating the growth of plants and trees that provide food and habitat for them. Beaver dams are associated with increased songbird diversity (Cooke and Zack, 2008). Shrubs increase, in the long run increasing moose and hare populations, and thus wolf populations. When wolves eat beaver, they excrete less intestinal parasites called cestodes (tapeworms are in this group), which are harmless to wolves, but can be picked up by moose, which they do harm.

Beaver dams can also create wetlands other than ponds that help some of the species discussed here as well as other species.

Both the number and percentage of keystone species appear much larger than commonly thought, indicating commensalism and symbiosis (since often the keystone species are aided by many of the species they help) account for a great amount of the tremendous biodiversity we observe on the planet.

Commensalism is ubiquitous. It is likely that every species is a commensal benefactor to at least one species and a commensal beneficiary to at least one species. From barnacles on sea turtles to epiphytes on trees, every large organism has one or more other species living on it.

References

Almeda, F. (2012). Personal communication.

von Beeren, C. & Tishechkin, A. K. (2017). Nymphister kronaueri von Beeren & Tishechkin sp. nov., an army ant-associated beetle species (Coleoptera: Histeridae: Haeteriinae) with an exceptional mechanism of phoresy. *BMC Zoology*, Volume 2, Article 3. https://doi.org/10.1186/s40850-016-0010-x

Pirt, S. J. & Lee, Y. K. (1983). Enhancement of methanogenesis by traces of oxygen in bacterial digestion of biomass. *FEMS Microbiol.* Letters 18 (1-2): 61-3.

Stamets, P. E., et al. (2018). Extracts of polypore mushroom mycelia reduce viruses in honey bees. *Scientific Reports* 8, Article no. 13936. https://doi.org/10.1038/s41598-018-32194-8

Reddy, G. V. P., & Guerrero, A. (May 2004). Interactions of insect pheromones and plant semiochemicals. *Trends in Plant Science* 9 (5): 253-61. https://doi.org/10.1016/j.tplants.2004.03.009

Yamazaki, K. (2007). Cicadas "dig wells" that are used by ants, wasps and beetles. *European Journ. of Entomol.* 104 (2): 347–9.

Mills, L. S., Soule, M. E., & Doak D. F. (1993) The keystone-species concept in ecology and conservation. *BioScience* 43 (4): 219-224.

Erwin, T. L. (1983). Tropical forest canopies: the last biotic frontier. *Bull. of the Entomol. Soc. of Amer.* 29: 14-9.

Gibling, M. R. & Davies, N. S. (2012). Palaeozoic landscapes shaped by plant evolution. *Nature Geoscience* 5: 99-105.

Clemente, C. J., Cooper, C. E., Withers, P. C., et al. The private life of echidnas: using accelerometry and GPS to examine field biomechanics and assess the ecological impact of a widespread, semi-fossorial monotreme. *Journ. Exp. Biol.* 219 (20): 3271–83. https://doi.org/10.1242/jeb.143867

Wilkinson, M. T., et al. (1 Dec. 2009). Breaking ground: Pedological, geological, and ecological implications of soil bioturbation. Earth-*Science Reviews* 97 (1): 257–72. Bibcode: 2009ESRv...97..257W. doi: 10.1016/j.earscirev.2009.09.005

Schwartzman, D. W. (1999). *Life, temperature, and the Earth: the self-organizing biosphere.* Columbia Univ. Press, Columbia, N. Y.

McIlroy, D. & Logan, G. A. (1999). The impact of bioturbation on infaunal ecology and evolution during the Proterozoic-Cambrian transition. *PALAIOS* 14 (1): 58–72. https://doi.org/10.2307/3515361

Braeckman, U., et al. (28 Jan. 2010). Role of macrofauna functional traits and density in biogeochemical fluxes and bioturbation. *Marine Ecol. Progress Series*. 399: 173–86. Bibcode: 2010MEPS..399..173B. doi: 10.3354/meps08336. ISSN 0171-8630

Pillay, D. (23 June 2010). Expanding the envelope: linking invertebrate bioturbators with micro-evolutionary change. *Marine Ecology Progress Series* 409: 301-3. Online pub. date: June 23, 2010. Print ISSN: 0171-8630. Online ISSN: 1616-1599. doi: https://doi.org/10.3354/meps08628

Meysman, F. J. R., Middelburg, J. J., Heip, C. H. R. (Dec. 2006). Bioturbation: a fresh look at Darwin's last idea. *Trends in Evol. and Ecol.* 21 (12): 688-95.

Grazhdankin, D. (March 2014). Patterns of evolution of the Ediacaran soft-bodied biota. *Journ. of Paleontology* 88 (2): 269-83. Pub. online by Cambridge Univ. Press: 15 Oct. 2015. doi: https://doi.org/10.1666/13-072

Remmers, W., et al. (23 Sept. 2016). Elephant (Loxodonta africana) footprints as habitat for aquatic macroinvertebrate communities in Kibale National Park, south west Uganda. *African Journ. of Ecol.* 55 (3): 342-35. https://doi.org/10.1111/aje.12358

Platt, S. G., et al. (9 Dec. 2019). *Water-filled Asian elephant tracks serve as breeding sites for anurans in Myanmar.* Pub. online. doi: https://doi.org/10.1515/mammalia-2017-0174

Kemp, P. S., et al. (2012). Qualitative and quantitative effects of reintroduced beavers on stream fish. *Fish and Fisheries* 13 (2): 158-81. doi: 10.1111/j.1467-2979.2011.00421

Stevens, E., Paszkowskia, C. A., & Foot, L. (Jan. 2007). Beaver (Castor canadensis) as a surrogate species for conserving anuran amphibians on boreal streams in Alberta, *Canada. Biol. Conservation* 134 (1): 1-13.

Cooke, H. A. & Zack, S. (2008). Influence of beaver dam density on riparian areas and riparian birds in shrubsteppe of Wyoming. *Western North Amer. Naturalist* 68 (3): 365-73. https://doi.org/10.3398/1527-0904(2008)68[365:IOBDDO]2.0.CO;2

CHAPTER 7. PHOTOSYNTHETIC ORGANISMS ARE AT THE BASE OF FOOD WEBS, ARE ECOSYSTEM ENGINEERS, AND AID AND ARE CLOSELY LINKED TO OTHER LIFE

Major contributions of plants and other photosynthetic organisms to bio-diversity not discussed in this chapter, such as carbon sequestration, providing habitat for nitrogen-fixing bacteria, providing carbohydrate to mutualistic fungi and helper bacteria, are discussed in other chapters.

Photosynthetic organisms include phytoplankton, plants (including trees), and some other organisms. The photosynthesis that these organisms carry out produces food in the form of carbohydrate used by these photosynthesizers and the herbivores that eat them. The carbohydrate they produce indirectly supports predators and decomposers as it is passed up the food webs when predators eat the herbivores and decomposers break down dead organisms and feces. Therefore, photosynthetic organisms, by converting sunlight to carbohydrate, support all of the above-ground food webs on land and most of the food webs in the sea. They are very important to life and biodiversity because they are at the base of these many food webs. Photosynthesizers also produce the oxygen needed by the vast majority of organisms on land and in the sea (see Chapter 4). Plants, especially trees, change ecosystems such that they support far higher biodiversity than there would be if they merely provided photosynthesis. Trees sustain forests, including themselves and all other life in the forests, by creating local rainfall and aiding cloud formation. One rainforest tree will release about 260 gallons of water per day. In the Amazon rainforest, at least half the rainfall comes from the trees, while the other half results from water evaporated and blown inland from the Atlantic Ocean (Salati, 1987). Rainforests affect Earth's reflectivity; this stabilizes wind currents and rainfall patterns. Rainforests stabilize climate in regions of the world far from them. Clearcutting rainforests around the equator has caused massive storms far from the cutting. For example, deforestation in the Amazon and Central America severely reduces rainfall in the lower U.S. Midwest during the spring and summer seasons and in the upper U.S. Midwest during the winter and spring, respectively, and tree removal in Southeast Asia greatly negatively impacts China and the Balkan Peninsula (Avissar and Werth, 2005).

Plants allow animal species to migrate and expand their ranges. Early humans were able to leave Africa and travel through northern Arabia to Eurasia as early as 300,000 to 500,000 years ago, only because it was covered in lush grassland as early as then. Stone tools and probable cut marks support these dates (Roberts et al., 2018). The presence of animal prey, also dependent on the grass, was also needed for this migration.

Plants and their mycorrhizal fungi create and build soil, and hold it in place, preventing erosion. Leaves and dead plants fertilize soil. Trees on the forest edge protect trees in the interior of the forest from wind and desiccation. Trees on islands protect soil from wind-blown salt. Trees stabilize river banks. Without trees, there would be no rivers, and water would flow in a fan-like fashion. Without rivers, diversity of aquatic invertebrates, fish, amphibians, and many other life forms would greatly diminish. The shade of trees keeps river water at temperatures favorably cool for aquatic life. *Forests filter and clean water and soil.* Not only do trees clean pollutants, the Chinese Brake Fern (*Pteris vittata*) cleans soil and water of arsenic, accumulating levels that would kill most organisms in specialized cells. Trees shade soil from the sun's heat, protecting it from drying into a hard, brick-like consistency. They provide shade that allows plants and young trees to thrive, and animals to escape the sun's heat. Plants, especially trees, are ecosystems that provide food and a three-dimensional habitat for many prokaryotes, fungi, epiphytes, invertebrates, and vertebrates. The habitat provided by living organisms that has the most species per unit area on land is the rainforest canopy. It was estimated that one hectare (about 2.47 acres) of seasonal Panamanian forest canopy may have over 41,000 species of arthropods (Erwin, 1982). The ecosystem in the ocean with the largest number of species is the coral reef. Coral is a symbiotic alliance between an animal and a photosynthetic alga. Together they provide a three-dimensional habitat for over 4,000 species of fish and a tremendous diversity of invertebrates. Coral reefs cover less than 0.1% of the world's oceans, about half the area of France, yet they support over 25% of all marine species (Spalding & Grenfell, 1997).

Plants caused the evolution of their leaves (Beerling 2001, 2007). When plants had roots but no leaves, they carried out photosynthesis with their

Photosynthetic Organisms Are at the Base
of Food Webs, Are Ecosystem Engineers,
and Aid and Are Closely Linked to Other Life

stems. Their roots and symbiotic root fungi weathered minerals. And their leaves and other debris accumulated on the soil, forming a moist, acidic environment that weathered soil minerals. The weathering sequestered tremendous quantities of carbon dioxide, greatly lowering its concentration in the air. Plants need carbon dioxide for photosynthesis, so there was strong selection pressure to photosynthesize more efficiently by maximizing surface area, favoring the evolution of leaves.

This was a big breakthrough and innovation that allowed plants to occupy a new adaptive zone with many new niches available to them. This led to tremendous plant diversification into many new species. This caused the diversification and creation of many new species of mammals, birds, reptiles, amphibians, insects, other arthropods, fungi, bacteria, viruses, and other groups that used the new plant species for food, habitat, and other services. Then the root, shoot, and leaf co-evolved together, each putting selective forces on the others to progress (Raven and Edwards, 2001). Once leaves, shoots, and roots evolved to a sufficient level of sophistication, competition between neighboring plants for light and to avoid being shaded out intensified. It is well documented in the fossil record that, in response to this competition, plants got taller and leaves became larger (Chaloner and Sheerin, 1979). This continued until forests of tall, leafy trees became a dominant feature of the land worldwide by 360 mya. This also caused a great diversification of tree species, which led to a tremendous diversification into many new species of animals, fungi, and other groups that used the trees for food, habitat, and other services.

Plants were major drivers of diversity of the insects that fed on them. An analysis found several species of plant-eating beetles had enhanced rates of diversification, indicating a series of diversifications into many new species spurred by coevolution with flowering plants, collectively representing almost half of the species of beetles (Farrell, 1998). This is significant because there are more species of beetle (order Coleoptera) than any other order of animal, about 1.5 million species. Ehrlich & Raven (1964) showed that coevolution between plants and the insects that eat them, primarily flowering plants and butterflies, resulted in great diversification of both groups, and hypothesized that insect diversity is intimately tied to the rise and diversification of flowering plants in the Cretaceous Period, which lasted from 145 to 66 mya.

Leaf litter provides nitrate for denitrifying bacteria, which remove excess nitrogen from the soil and return it to the air. Nitrogen is a nutrient, but is harmful in quantities that are too great.

All plants, especially trees, provide, a three-dimensional habitat, often with platforms, for a multitude of species of squirrels, bats, birds, lizards, insects, other invertebrates, fungi, plants that grow on trees, unicellular eukaryotes, prokaryotes, and viruses.

Plants make up 83% of the total biomass of all life on Earth.

References

Salati, E. (1987). The forest and the hydrological cycle, in *The Geophysiology of Amazonia*, ed. by Dickinson, R. E. pp. 273– 96, John Wiley, New York, N. Y.

Avissar, D. & Werth, D. (April, 2005). Global hydroclimatoloigal teleconnections resulting from tropical deforestation. *Journ. of Hydrometeorology* 6: 134-45.

Roberts, P., et al. (29 Oct., 2018). Fossil herbivore stable isotopes reveal middle Pleistocene hominin palaeoenvironment in 'Green Arabia'. *Nature Ecol. & Evolution* 2: 1871–8.

Erwin, T. (1982). Tropical forests: Their richness in Coleoptera and other arthropod species. *The Coleopterists Bulletin* 36 (1): 74-5.

Spalding, M. D. & Grenfell, A. M. (1997). New estimates of global and regional coral reef areas. *Coral Reefs* 16 (4): 225–30. doi: 10.1007/s003380050078

Beerling, D. J., et al. (2001). Evolution of leaf-form in land plants linked to atmospheric CO2 decline in the Late Palaeozoic era. *Nature* 410: 352-4.

Beerling, D. (2007). *The Emerald Planet: How Plants Changed Earth's History*. Oxford University Press, Oxford, U. K.

Raven, J. A. & Edwards, D. (2001) Roots: Evolutionary origins and biogeochemical significance. *Journ. of Exper. Botany* 52: 381-401.

Chaloner, W. G. & Sheerin, A. (1979). Devonian microfloras. *Special papers in Palaeontol.* 23: 145-61.

Farrell, B. D. (24 July 24 1998). "Inordinate Fondness" explained: why are there so many beetles? *Science* 281 (5376): 555-9. doi: 10.1126/science.281.5376.555

Ehrlich, P. R. & Raven, P. H. (Dec. 1964). Butterflies and plants: A study in coevolution. *Evolution* 18 (4): 586-608.

CHAPTER 8. HERBIVORES GENERATE BIODIVERSITY

Herbivores increase biodiversity, here meaning simply the number of species, in the following ways. They eat large quantities of plants, controlling their populations. For example, each adult wild elephant eats between 149 and 169 kilograms (330 and 375 pounds) of vegetation per day. Herds usually have about 10 to 20 elephants. Sometimes several herds merge into super herds of 100 or more. If plants were not regulated by herbivores, they would grow to be so dense that in some ecosystems, some animals would not be able to move through them. They would deplete their soil, minerals, and water, and die off. Herbivores selectively eat plants that are the better competitors, preventing them from driving the plant species that are poorer competitors extinct in their ecosystem. They keep plants from reaching their carrying capacity, a point where better competitors drive poorer competitors extinct. Large herbivores from the Pleistocene Epoch (about 2.58 million to 11,700 years ago), such as mammoths, and living herbivores, such as elephants and bison, maintain mosaics of *both* grasslands and forests by destroying trees, increasing the number of habitat types and thus the number of species. Without these large herbivores, there would have been and would be only forests and no grasslands in the areas where these herbivores lived and live. Herbivores ensure that fires are not too numerous nor too severe by regulating the density of plants that fires need for fuel. Large herbivores till and aerate the soil with their hooves by walking on it and running from their predators, allowing plants to thrive.

The hippopotamus enhances wetland habitat, making paths through high grasses. Large herbivores such as elephants and hippopotami add and distribute nitrogen and other nutrients to the soil by their movement and defecation. Herbivores, especially insect herbivores, are an important food source for predators; this supports many food webs on land. Other herbivores support predators and food webs in fresh water and the sea. Coevolution of herbivores and plants increases diversity of both. Plants evolve chemicals to guard against herbivores, and herbivores evolve defenses against these toxins. Caterpillars and butterflies even incorporate the plant toxins into their bodies for protection against their predators. They store the toxins in the peripheral areas of their bodies, thus avoiding the damaging effects of the toxins. Once they have the toxins in their bodies protecting them, they evolve bright, contrasting colors that warn their predators that they are toxic. In the sea, herbivores control phytoplankton, with the result that these photosynthesizers do not deplete their nutrient supply and thus have population crashes. Herbivores increase species diversity of the plant species they eat by providing an intermediate amount of disturbance. Eating too much would decrease diversity of the photosynthesizers that they eat, but if they were not present or only ate only small amounts of photosynthesizers, the better competitors would cause the poorer competitors to become locally extinct. Intermediate herbivory prevents this and keeps the number of plant or other photosynthetic species high. Herbivores generally tend to consume at reasonably intermediate levels because they are controlled by predators and disease by negative feedback, keeping their numbers balanced at an intermediate level. Lubchenco (1978) studied the effects of the snail called the common periwinkle (*Littorina littoria*), on the diversity of algae in tide pools on the New England coast. Pools with low snail density had as little as one species of alga, the green alga called sea lettuce (*Uva intestinalis*), which is the best competitor and which eliminated all other algal species. In pools with intermediate snail density, there was the highest algal diversity, with sometimes as many as ten or more algal species. In pools with very high snail density, the snails ate all the species of algae except the inedible red alga, Irish moss (*Condrus crispus*), and encrusting coralline algae. Some coral reef herbivores, such as fish and snails, increase reef diversity by controlling the growth of the algae that they eat, preventing the algae from suffocating the coral. Since the plants that herbivores eat benefit from these herbivores, herbivory is a form of symbiosis.

References

Lubchenco, J. (1978). Plant species diversity in a marine intertidal community: importance of herbivore food preference and algal competitive abilities. *The Amer. Naturalist* 112 (983): 23-39. https://doi.org/10.1086/283250

CHAPTER 9. PREDATORS AND PARASITES GREATLY INCREASE BIODIVERSITY

Predators increase and maintain the biodiversity of their prey and health of their ecosystem by several mechanisms. They prevent collapses of their prey population and destruction of their ecosystems by regulating their prey numbers, so their prey do not become over-populated and destroy their environment. Without predators, their prey would eat their plant or animal food to such low levels that they would undergo a population crash or even a local extinction. Predators keep their prey populations stable because they increase in number and eat more prey when prey populations are high and decrease in population size and eat less prey when prey populations are low. They also eat more prey per predator when prey populations are high, and less per predator when prey populations are low. Thus, they stabilize prey populations by negative feedback, and increase the chance prey populations will not go locally extinct.

Predators maintain a high number of their prey species by eating the most successful competitors. Paine (1966) demonstrated this by studying the effects of the purple sea star, the main predator on the rocky shore habitat in Makah Bay, on the coast of Washington state. This sea star eats rock barnacles, gooseneck barnacles, mussels, limpets, and chitons. Paine removed all sea stars in one area, keeping it free of them, while leaving a control area undisturbed. He found after the sea star was removed, the experimental area declined rapidly from 15 to 8 species, while the control area maintained its diversity. The decline in the area where sea stars were removed was mainly because barnacles and California mussels, the best competitors, out-competed and eliminated the other species. Where the sea star was left in place, it preferentially ate these species, controlling their numbers, allowing the other species to persist, keeping the species diversity high. Paine (1966; 1974) also showed that the lower portions of mussel beds smother seaweed and prevent it from growing there. But if sea stars are present, they eat the mussels at the bottom of the mussel bed, allowing the seaweed to grow, which adds diversity to the system. The seaweed is habitat for many fish and invertebrate species. The mussels provide habitat to several species too, but the presence of both mussels and seaweed provides habitat to a greater number of species.

Predators also keep prey populations from reaching carrying capacity, so competing prey cannot reach a point where better competitors drive poorer competitors locally extinct. Predators cause their hooved prey to clump together, and flee when the predators appear, tilling and aerating the soil with their hooves. This improves the soil and allows plants to thrive maintain high species numbers. Predators cause their prey to move around and regulate prey numbers with the result that prey do not over-consume their food supply in local areas. For example, wolves cause deer and elk to eat for a while in a forest, then move to eat in another area, with the result that wolves cannot predict where their prey will be as easily. This keeps the deer and elk from depleting the trees and vegetation in forests locally. The wolf also controls deer and elk numbers, further protecting the forests. When the wolf was removed from the Greater Yellowstone ecosystem, forests were destroyed by over-eating by deer and elk, and populations of species of birds, reptiles, amphibians, insects, and other animals that are dependent on the forests crashed. Beavers declined and myriad species dependent on beaver ponds disappeared. Erosion of the soil and river banks, leading to siltation of rivers, occurred when the area was deforested by deer and elk, since trees hold the soil in place, preventing erosion. Hence, wolves indirectly prevent erosion and pollution of the rivers. A catastrophic crash of an ecosystem because a key predator is removed, such as the case just described, is called a *trophic cascade*. When wolves were re-introduced to this ecosystem, the other species returned and the biodiversity was re-established. In addition to wolves, keystone predators whose local extinction leads to collapses of ecosystems include mountain lions, coyotes, bats, and sharks. Predators strengthen prey populations by the culling the sick, old, and unfit. They reduce disease in their prey by eating the sick individuals. Predators coevolve with their prey, resulting in key innovations in both predator and prey. For example, the shells of clams, oysters, and mussels evolved as a defense against predation, and some predatory snails evolved chemicals to bore through these shells. These adaptations are often followed by large diversifications into many new species. This was the case after the shells of clams, oysters, and mussels evolved.

Parasites, like predators, regulate host populations and selectively kill the better competitor, increasing host diversity. They coevolve with their hosts, resulting in key innovations and diversification in both host and parasite. Viruses greatly help the evolution of their hosts because their hosts frequently incorporate their DNA, often resulting in large evolutionary innovations followed by tremendous diversification. Predator-prey and parasite-host relationships are in reality a form of symbiosis, when one considers that the prey and host benefit from their predator or parasite at the level of the population.

References

Paine, R. T. (1966). Food web complexity and species diversity. *American Naturalist* 100: 65–75.

Paine, R. T. (1974). Intertidal community structure: experimental studies on the relationship between a dominant competitor and its principal predator. *Oecologia* 15: 93–120.

CHAPTER 10. DECOMPOSERS ARE
INDISPENSABLE TO THEIR ECOSYSTEMS

Decomposers are essential to ecosystem health, and create very high biodiversity. They break down dead plants, animal carcasses, and feces. Without decomposers, huge piles of dead organisms and feces would build up on the ground, severely impeding the movement of animals. Additionally, carcasses and feces would build up, and disease-causing bacteria would grow on them. The amount of disease in wildlife would increase significantly. Decomposers recycle nutrients, such as carbon, nitrogen, oxygen, and phosphorus, making them available to the ecosystem. Without decomposers, ecosystems would run out of nutrients, and organisms would die as a result. Decomposers close the loop of energy flow, recycling nutrients from all levels of the food web, and providing them to primary producers such as plants and phytoplankton.

Decomposers represent a significant portion of the various food webs on Earth and are in all major groups of organisms. On land, they include bacteria, fungi, worms, slugs, snails, termites, bark beetles, woodboring beetles, ants, vultures, and hyenas. In the ocean, decomposers include bacteria, archaea, crabs, lobsters, hagfish, and sleeper sharks.

The Chernobyl nuclear accident resulted in radiation that killed a large portion of the decomposers of the forest there. This resulted in leaf litter accumulating; leaf litter decomposed at a rate 40% lower than the normal rate (Mousseau et al., 2014). The accumulation of leaf litter and dead trees increased the probability of abnormally large forest fires that could scorch the soil. Nutrients were not efficiently recycled to the soil and the ecosystem. Trees grew much slower than normal, and this is likely because of the lack of soil nutrients. As time passed, the thickness of leaf litter and dead trees on the forest floor increased in proportion to the level of radiation the area was exposed to, because of the loss of decomposers. The higher the radiation exposure, the greater was the loss of decomposers.

Decomposers keep diversity high within their trophic level because one decomposer species gets displaced by another in a consistent, sequential order. For example, In North America, turkey vultures are the first birds to arrive when a carcass appears, because they have the best sense of smell of the scavenger birds that occur there. Black vultures observe turkey vultures to find carcasses, and arrive next. Being more aggressive, black vultures displace turkey vultures. In some parts of desert ecosystems, caracaras arrive last, and displace both vulture species. The species that is better at finding the dead animals is the weakest competitor, and the strongest competitor is the slowest to get to the carcasses. This mechanism allows all decomposer species to persist in the community, and allows decomposers to have high biodiversity. This is a general principle applies that applies to many decomposer communities, such as ones in Africa with vultures, jackals, and hyenas that sequentially eat what is left of the carcass after the big predators such as lions and leopards who killed it are finished with it. After the animal decomposers are finished, the decomposition of the carcass continues with many species of fungi and bacteria getting nutrient from the carcass in a microbial succession that also keeps the diversity of decomposer species high.

References

Mousseau, T. A., et al. (May 2014). Highly reduced mass loss rates and increased litter layer in radioactively contaminated areas. **Oecologica** 175 (1): 429–37.

CHAPTER 11. GENOMES ARE CREATIVE GENETIC ENGINEERS WHOSE BEHAVIORS PROMOTE GENETIC VARIABILITY, EVOLUTION, LARGE ADAPTIVE EVOLUTIONARY INNOVATIONS, AND DIVERSIFICATION INTO NEW SPECIES AND OTHER TAXA

The genomes of organisms perform genetic engineering, and their behavior causes increased genetic variability within populations, rapid evolution, large evolutionary breakthroughs, and, as a result of the latter, diversification into many new species and hence increased biodiversity. This sometimes causes increases in DNA content and complexity of the genome. Genomes of organisms have a tendency to evolve new forms and to diversify. Surprisingly large numbers of evolutionary breakthroughs and diversifications are in part driven internally, by the behavior of the genome itself. Watson and Szathmáry (2016) even posit that evolution can learn from past experience to better evolve; this too involves behavior of the genome. Rearrangement of the genome is thought to be a driving force in evolution, since it gives rise to novel gene combinations (Aguilera and Gomez-Gonzales, 2008). There are several mechanisms by which genomes function to increase variability and diversity, and to promote adaptive evolutionary change. Genomes of all sizes are a patchwork of genetic material coming from many different kinds of organisms and species with different histories. This includes the human genome.

There are two important terms to understand for this chapter. The *phenotype* consists of characteristics and traits of an organism that can be seen. The term covers the organism's physical form and structure, developmental processes, biochemical and physiological properties, and behavior. The phenotype includes the outward appearance (shape, structure, coloration, pattern, size, weight), and the form and structure of the internal parts like bones and organs. Phenotype contrasts with *genotype*, which is the complete set of genetic material, the entire genome, of an organism. The adjectives are phenotypic and genotypic.

Natural Selection Optimizes
the Amount and Nature of Variability

Natural selection, in addition to acting directly on traits of the organism, acts on and optimizes the amount and nature of genetic and phenotypic variability. Genetic and phenotypic changes need to be compatible with the integrated, well-adapted genome, organism, and population to maximize the probability that they will survive and be passed on, or bring about adaptive evolutionary innovations. The amount of variability is selected primarily to correspond to the variability of the environment. As to the nature of variability, it has long been believed that mutations are random with respect to selection. However, a key mechanism by which natural selection optimizes the amount and nature of variability is that *mutations are not random with respect to natural selection, and adaptive mutations are more likely to occur than expected if such randomness were the case.*

There are studies discussing or showing this idea that the mutation rate itself is subject to natural selection (for example, Kimura, 1967; Lynch, 2010; Zhang, 2022). Natural selection favors the lowering of mutation rates, since mutations are mostly deleterious (Kimura 1967; Drake 1991; Dawson 1998; Lynch 2010). Swings et al. (2017) showed that mutation rates in the human intestinal bacterium *E. coli* rise when cells experience higher stress. Melamed et al. (2022) showed mutations for resistance to malaria are more common in people in Africa where malaria occurs, than in people from Europe, where it does not. This difference is due to natural selection.

Most mutations are deleterious because they disrupt the well-integrated, coordinated phenotype and genotype, which are well-adapted to their environment due to eons of natural selection. If one randomly changes any part of a car, it is likely to make the car run more poorly, because the car was already well built for the road and the new part is unlikely to be compatible with the rest of the car. Similarly, a random mutation will likely disrupt the coordinated, integrated genome and phenotype of the organism. Random point mutations are especially likely to be deleterious because they are very likely to disrupt the coordinated genetic system. A point mutation is a mutation that affects only one base or a few bases that make up DNA. Larger mutations, if random, are the most likely to be disruptive and hence deleterious, since they affect a large portion of the genome. However, *segments of DNA that have been favored by selection for long time periods, are large enough to be compatible with larger genetic systems, and that have evolved with genetic systems long enough to be compatible with other genetic systems have a significantly higher probability of being adaptive and helpful to the organism and genetic system they are added to than small, random mutations. Because of they have been selected for their fitness and for their compatibility with genetic systems, they are not fully random with respect to natural selection. And being somewhat large at times, they have a higher likelihood of causing a large adaptive change.* They still have a much higher probability than 50% of being deleterious because any change to a well-adapted and self-compatible system is likely to disrupt it. But they have a much higher probability of bringing about adaptive change than random mutations. This is what I mean by natural selection acts on the nature of variability. The following discussion is about these kinds of genetic changes, which are more likely to be compatible with the genetic systems they integrate with, and hence adaptive.

Sexual Reproduction Causes Compatible, Adaptive Genetic Change with High Probability

Sexual reproduction is a first mechanism by which the genome's behavior promotes diversity while maintaining the compatibility of the parts of the genome to each other. It is among the most common ways that variability, adaptive macroevolutionary breakthroughs, and diversification result from the genome's behavior. It keeps populations from going extinct by keeping pace with environmental changes. It aids the speciation process. And the formation of new species by hybridization between species is one of a number of mechanisms by which sexual reproduction can increase diversity.

Sexual reproduction results in a loss of half the genome compared to asexual reproduction. It has other costs, such as the need to find and compete for suitable mates. There are several hypotheses attempting to explain why it evolved given these high costs. None have been shown to make up for the cost. I will now present a novel hypothesis to explain why it evolved. I my view, it is probable that the solution lies in a combination of causes, the most important one being my explanation, which follows.

Organisms need genetic variability and changes in their genetics in order to change and adapt as their environment changes through time. The environment is constantly changing due to evolution of predators, changes in weather, and so on. Genetic variability also allows organisms to adapt to and use differing areas in their environment. Variability is necessary, too, for large adaptive breakthroughs in evolution, speciation, and diversification. But most mutations are deleterious and disrupt the integration of the organism. Sexual reproduction allows genetic variability and new gene combinations for the production of novel forms that natural selection can act on, and *does so with a much lower probability of disrupting the integration and compatibility of the genome.*

This is because in sexual recombination, entire chromosomes or large segments of chromosomes that are already compatible with the rest of the genome are exchanged between chromosomes, resulting in a new chromosome well integrated and compatible with the rest of the genotype. (A chromosome is a structure found inside the nucleus of a eukaryotic cell, composed of most of the cell's DNA tightly coiled around proteins; it carries most of the genetic material of the cell. In prokaryotes, which do not have a nucleus, the DNA scattered throughout the cell can be called the chromosome). These chromosomes are well-adapted DNA segments that have experienced long periods of natural selection in the environment of an integrated genetic system. They coevolved with the rest of the genome and are compatible with it. The chromosomal segments or entire chromosomes involved in sexual recombination are as a rule kept in their original form, the chromosomes that exchange the DNA are very similar to each other, and the DNA segments that are exchanged usually move into exactly or approximately the same part of the chromosome from whence they came--that is, there is generally no change in the location of the exchanged segments relative to the other genes and parts of the chromosome they end up in. The DNA segments that are exchanged are compatible with the regions of DNA they move into because they coevolved with regions of DNA similar to the regions they land in. Thus, the basic structure, fitness, integration, and self-compatibility of the genetic system is preserved. Therefore, sexual reproduction results in a relatively high probability of adaptive change. This contrasts with a random mutation, which provides variability for natural selection to act on, but which is much more likely to disrupt the integration and self-compatibility of the genetic system and be unfavorable to the organism.

This does not mean all novel combinations of genes produced by sex will be adaptive and compatible.

About 10-20% of clinically recognized pregnancies in humans end in miscarriages, which are spontaneous abortions, and this number is likely higher because many early miscarriages go undetected before a woman knows she is pregnant. A large percent of spontaneous abortions occur as a result of chromosomal abnormalities in the fetus. It is reasonable to assume that many of these are due to maladaptive chromosomal and gene combinations due to sexual recombination. However, *the fact that the exchange is between entire chromosomes and large segments of chromosomes that have coevolved to be compatible for long time periods greatly increases the probability that the genetic combinations will be compatible and adaptive, and even produce novel adaptive traits. This is a principal reason, and perhaps the principal reason, that sexual reproduction was favored by selection.* The fact that sexual reproduction is common supports my thesis that there is selection for an optimal amount and type of variability.

Enzymes evolved that induce crossing over, the exchange of segments of DNA between paired chromosomes during sexual recombination. This shows that crossing over and hence sex were selected for, further evidence that natural selection optimizes the amount of variability and favors variability that is compatible with the genome.

Sexual reproduction also produces new combinations of genes that can lead to macroevolutionary breakthroughs. In a hypothetical example, a fish could have a gene complex that produces legs instead of fins and another fish could have a gene complex for breathing air. When sex brings the two gene complexes together, an organism better adapted to land with both adaptations appears as an advance toward the evolution of amphibians.

Sexual recombination can also cause the duplication of genes by unequal crossing over. One chromosome in the cross over loses a gene, and the other ends up with an additional gene. Thus, one copy of the gene can carry out the gene's function while the other copy, called the *pseudogene*, can mutate and evolve without expressing itself. The functional gene is dominant and masks the effects on the phenotype of the pseudogene. The pseudogene can randomly mutate until by chance it mutates to a novel adaptive form that produces a fit phenotype. This could be a large adaptive change. This adaptation would then spread through the population. This is a mechanism by which a large, adaptive evolutionary change can occur in a very short time, the punctuation in punctuated equilibrium. This is also a mechanism by which sex causes an increase in the amount of DNA in a species.

Amazingly, rates of recombination vary tremendously between different regions in the human genome, suggesting selection for more variability in some areas of the genome than others (Lander et al. 2001). There are recombination hotspots in the genome, where rates of recombination can be hundreds of times more than in the surrounding region (Jeffreys et al., 2001). There are more than 30,000 recombination hotspots within the human genome (Baudat et al., 2010) where the average number of recombination events per hotspot is one per 1,300 sexual cell divisions, which is far higher than average (Myers et al., 2006). It is astonishing that natural selection has created some areas on chromosomes where recombination is much more frequent than in other areas. Clearly, it is advantageous to recombine and thus generate variability more in some areas of the genome than others. This is yet more evidence that natural selection optimizes the amount of genetic variability in organisms.

There are many other hypotheses attempting to explain the predominance of sexual reproduction in multicellular organisms. The complete explanation may be a combination of a number of the hypotheses. But I am confident that my hypothesis is at least part of the explanation.

Viruses Were Important in Promoting Major Macroevolutionary Breakthroughs in All Cellular Organisms

Cellular organisms are all organisms with cells, from bacteria to plants to animals; they consist of all organisms that are not viruses. Viruses generally undergo a historical pattern with their hosts. First, when a new virus appears, it causes disease and often kills some of its hosts. Then the hosts evolve resistance to the virus. The virus evolves to be less virulent, because it needs its host to survive, since its host is its habitat. And allowing its host to live a longer life gives the virus more time to infect a new host. This selection to be gentle on the host is even more pronounced after the virus has been around a while, because there are less hosts because the virus kills a portion of them. Thus, *the virus causes conditions for selection for it to be less disease-causing.* In the final step, the virus incorporates its DNA into the host DNA, becoming part of the host, and symbiotic with it, no longer harming the host at all. This virtually gives the virus immortality. This is significant because incorporation into the host DNA sometimes gives the host novel, adaptive genes, causing evolutionary breakthroughs. These usually are regulatory genes. Sometimes viruses donate entire networks of regulatory genes to their host, giving the host many regulatory genes that work together. This can cause large adaptive evolutionary transitions in the host. *This is evolutionary symbiosis. All the evolutionary breakthroughs that viruses cause in their hosts, many of which I will now discuss, are the result of symbiosis, with the virus receiving a habitat from the host, and host receiving viral DNA, which helps its evolution. This is support for the thesis of Chapter 5, that symbiosis is fundamental and greatly enhances biodiversity; this is a basic tenet of the Pachamama Hypothesis.*

Viruses can break free of the host genome they are integrated into. Sometimes they take some of their host's DNA with them when they leave the host DNA. They can then integrate into another host's DNA, giving their new host new DNA for use. This is called *transduction*, and can occur between hosts of the same or different species. It is a source of variability, novel DNA, and increase in DNA content. It is likely important in increasing species diversity, evolutionary rates, and evolutionary breakthroughs. Transfer of DNA by viruses by transduction is particularly common in bacteria (Canchaya et al., 2003), and promotes prokaryotic diversity. But it also occurs in eukaryotes. Anderson (1970) summarized the extensive evidence that viral transduction is a key mechanism for transporting segments of DNA across species and phylum barriers, and that much evolution depends largely on this transfer. Viruses are a key natural mechanism for transferring genes between species (Liu et al., 2010). *It is thought they played a central role in very early evolution, before the splitting of life into the three domains of bacteria, archaea, and eukaryotes, and at the time of the last universal common ancestor of all cellular life on Earth.*

Transduction is a form of symbiotic genetic coevolution. The virus benefits by living off the host, while its host can potentially benefit in its evolution by acquiring novel DNA that is sometimes adaptive.

About 8% of the human genome was built from viruses (Lander et al., 2001). A large percentage of the DNA of other mammals and many other taxa also came from viruses.

There are now several hypotheses that put viruses at center stage for major evolutionary transitions and breakthroughs in cellular organisms (Claverie, 2006, and references therein). This was accomplished by viruses giving genes to their hosts. Viral agents play critical roles within the roots and stem of the tree of life (Villarreal and Witzany, 2010). Viral proteins replaced cellular proteins during the early evolution of the nuclear DNA replication (DNA duplication) system in prokaryotes (ibid.), and viruses played a key role in the origin of DNA replication mechanisms in cellular organisms (ibid.). In fact, Filée et al. (2003) showed numerous transfers of genes between viruses and cells, in both directions, in the genes for four proteins involved in the replication of DNA, and the manufacture of molecules used to make DNA. This means *at least some of the genes for the manufacture of the molecules used to make DNA and for the replication of DNA in fungi, plants, and animals came from viruses.* They suggested that the transfer of genes from viruses to the cells of organisms, and the replacements of the genes of cells by the genes of viruses are important sources of new genetic material in the evolution of organisms, and that *viruses are major players in the evolution of the genomes of organisms.* Also, transcription of DNA into RNA to code for proteins, repair of DNA, and fine-tuned regulation of genes are now known to also be of viral origin (Villarreal, 2005). Furthermore, viruses might have helped the evolution of cells, the origin of DNA as the genetic material, increase in prokaryote and eukaryote genome size and complexity, the evolution of the nucleus of the cell, and an increase in the diversity of life (Koonin, 2016). These ideas will be now discussed in more detail.

Szathmáry and Demeter (1987) and Szathmáry and Maynard Smith (1997) showed that parasites, including viruses, promote natural selection between entire groups of cooperating replicating segments of DNA, which can lead to the evolution of cells. Of course, other factors than viruses helped the evolution of cells as well. Takeuchi et al. (2011) showed that a population of replicating segments of DNA that were in a uniform area, before cells evolved, is prone to be doomed to collapse from parasites, and the only way for such a population to be stable is compartmentalization, which would occur by the appearance of cells. The replicating DNA segments would be separated into cells, and the cell membranes would separate the DNA segments from each other, making it more difficult for a parasite such as a virus to spread from one DNA segment to another. Thus, this slows the rate of transfer of parasites to new hosts by creating barriers to their spread, and creates conditions for cooperation between replicating segments of DNA of the host because they would be in the same cell and they would all benefit from its survival (Higgs and Lehman, 2015; Shay et al., 2015). Thus, *viruses seem to have played a key role in the evolution of the first cells.*

Takahashi and Marmur (1963) posited that viruses provided the selective pressure for DNA to replace RNA as the genetic material. Proteins can act as enzymes. Enzymes carry out the chemical reactions organisms need to survive, grow, and reproduce. DNA acts as the genetic material of the cell. But proteins cannot act as the genetic material and DNA cannot act as an enzyme. Only RNA can act as both genetic material and enzymes. For this reason, it is thought (and reasonable to assume) that RNA was the first biological molecule, preceding DNA, and hence was the first genetic material (Atkins et al., 2010; Bernhardt, 2012). This is the RNA world hypothesis (Gesteland et al., 2006). Some viruses still use RNA as their genetic material. One reason RNA was replaced by DNA as the genetic material is because DNA is more stable and harder to degrade than RNA. DNA is now the genetic material in all organisms except some viruses. Mathematical modelling indicates that defense against parasites like viruses is most effective when the genetic replicator (such as DNA and RNA) and catalyst (protein that helps the genetic replicator duplicate itself) are separated into two different types of molecules with specialized systems, such as DNA and proteins (Takeuchi et al., 2011). RNA can act as both replicator and catalyst, causing the two to not be separated, but to be one molecule. But DNA cannot be the catalyst, so with DNA as the replicator (genetic material), there are the two separate molecules: DNA as replicator, protein as catalyst. So defense against viruses is more effective. *This indicates that the coevolution of prokaryotes and bacterial viruses aided the evolution of DNA as the genetic material.* Thus, *viruses were a key selective agent in DNA replacing RNA as the genetic material in early evolution.* However, the replacement of RNA by DNA was also selected for because it conferred other advantages. For example, it was favored by natural selection because DNA has greater stability, and double-strandedness, which allows the ability for DNA repair (ibid.). Thus, phage likely played an important role, but other selective factors were probably involved as well. Furthermore, the replacement of RNA with DNA allowed the evolution of larger, more complex genomes (ibid.) because of the increased stability of DNA compared to RNA. The larger, more complex genomes allowed for large, adaptive evolutionary transitions to new forms, which caused entry into new adaptive zones with many open niches, allowing diversifications into great numbers of new species. RNA genomes would likely not have been stable enough to allow these larger genomes, this tremendous complexity, these macroevolutionary advances, or the great diversity, to occur. Furthermore, it has been suggested that the transition from RNA to DNA genomes occurred in the viral world, and that cellular DNA and its replication machineries originated via transfers of DNA from DNA viruses to RNA cells (Forterre, 2006 and references therein).

Martin and Koonin (2006) proposed that viruses selected for the nucleus of the eukaryotic cell. The nucleus is the part of the cell that contains and encloses the DNA in eukaryotes. Bell (2001, 2006) proposed that the nucleus of the cells of eukaryotes evolved from a complex DNA virus. Koonin and Krupovic (2020) showed that large bacteriophages protect themselves from enzymes made by the bacterial host cell to attack them by encapsulating their genome inside the cell in a shell that is like a "nucleus" and that is impenetrable by some of these enzymes. The giant phage 201Phi2-1 forms a structure similar to a nucleus when it infects a bacterial cell (Chaikeeratisak et al, 2017). These last two observations suggest a possible viral origin of the cell's nucleus.

Witzany (2008) posited a viral origin of telomeres and telomerase, key elements of chromosomal structure and cell replication in eukaryotes. Telomeres are tiny caps at the ends of the DNA molecules that make up chromosomes. They protect the ends of chromosomes and keep them from fraying or sticking to each other, functioning much like the plastic tips on the ends of shoelaces. Chromosomes are structures in the nucleus of the cells of eukaryotes that contain the DNA as well as proteins associated with the DNA. Telomerase is an enzyme that adds DNA to telomeres to maintain their length. Without telomerase, telomeres and the chromosomes that they are on would gradually get shorter and shorter. This would be very bad for the health and eventually lethal for the organism that contains them.

Thus, *viruses contributed to the origin of eukaryotic cells!* This was a major evolutionary breakthrough. Eukaryotic cells are on average about 1,000 times bigger by volume than prokaryotic ones; contain an inner membrane system with organelles such as the mitochondrion and the nucleus surrounded by membranes; and are fully compartmentalized so that movement of molecules between compartments is greatly limited. There would have been spectacular diversification into very many new species after the appearance of the eukaryotic cell. The many new species provided habitats and niches for other species that evolved to use these habitats and niches, further increasing the number of species on Earth.

Forterre (2006) claims ancient viruses were at the origin of the three domains of life. These are the three biggest groups of life other than viruses. The three domains of life are archaea, bacteria, and eukaryotes. Bacteria and archaea are simple organisms that do not have a nucleus, mitochondria, or chloroplasts in their cells. This would be amazing if viruses gave rise to the three major groups of cellular life! He proposed that three independent transfers from DNA viruses to what were originally cells with RNA as the genetic material were at the origin of archaea, bacteria, and eukaryotes. This may explain why there are three discrete domains of life as distinct lineages and three types of modern cells, rather than a gradual gradation from, say, bacteria to eukaryotes; why there are three different kinds of protein-making factories (called ribosomes), one in each domain; and why each domain has its own specific DNA replication machinery. However, there may be other explanations for these observations. For example, ancestors that were intermediate between archaea and bacteria, or between eukaryotes and bacteria or archaea could have gone extinct long ago without leaving a fossil record. So the exciting idea has evidence supporting it, but is not yet proven. *Plasmids* are circular segments of DNA that are separate from the chromosome of prokaryote cells. Forterre views plasmids as transitional forms between DNA viruses and cellular chromosomes.

Viruses were important in the evolution from one-celled to multicellular organisms by providing natural selection favoring multicellularity and providing DNA needed for it. Multicellularity evolved multiple times in eukaryotes, and four times in complex ones with specialized cells and tissues: in animals, plants, fungi, and brown algae. It evolved in some prokaryotes, including some cyanobacteria (Schirrmeister, 2011). Multicellularity is a huge evolutionary breakthrough, transition, and a key evolutionary innovation. It allows an increase in size, complexity, lifespan, and the kinds of actions that can be done. It opened up a new adaptive zone, and hence a myriad of new niches. As a result, as soon as it evolved, multicellular forms underwent tremendous diversification into many new species that then filled these new niches. These new multicellular species provided many new habitats and niches that other species evolved to use, further increasing the number of species on Earth.

Programmed cell death (PCD) is part of and fundamentally linked to multicellularity. It is the death of a cell as a result of events inside of a cell, programmed by the cell. It is generally good for the organism. It controls pathogens and cell proliferation, and is part of normal development in multicellular eukaryotes. For example, fingers and toes cannot develop properly in a human embryo without PCD. Iranzo et al. (2014) have a mathematical model showing that for unicellular organisms, including some bacteria, under high virus load and imperfect immunity, joint evolution of the aggregation of cells and PCD is the optimal evolutionary strategy. Infected bacterial cells evolved to altruistically cause their own deaths, preventing other cells near them from being infected. This benefits the group—the aggregate of cells—at the expense of the altruistic cells that commit suicide. A model based on natural selection between entire groups with the groups with the most altruists cooperating and so defeating selfish groups that do not have internal cooperation cannot work because cheating cells that do not commit suicide would be selected for over the altruists that undergo PCD. The Iranzo et al. model works because many bacterial colonies are genetically identical clones, so the cells around the suicidal cell are genetically identical to it. So the altruistic cells that commit suicide are passing on their genes because they are helping cells next to them that are genetically exactly the same as they are. This is very Darwinian. It is natural selection favoring the cell that commits suicide because it helps other cells that share its genes. So it passes on its genes to the next generation through the cells next to it that have exactly the same genes as it does, and it so is an evolutionary success. This phenomenon of an organism passing on its genes by helping relatives that share its genes is called kin selection. The model of the linking of PCD and kin selection requires multicellularity. A single cell cannot help the development of a group of cells that are separate from it. On the other hand, if the cell is part of a multicellular organism and commits suicide, it can help, for example, fingers separate and develop into a hand. Because the organism is multicellular, the cellular suicide helps nearby cells and the organism it is part of. Therefore, the model shows that multicellularity and PCD are fundamentally linked, that one cannot evolve without the other. Multicellularity and PCD are also linked to kin selection, and the system only works if the cells are closely related genetically, as is the case for cells of the same organism, where they are identical genetically. The model also implies that viruses were important in the evolution of multicellularity because PCD is linked to multicellularity, and PCD is crucial in fighting viruses. In fact, Quistad et al. (2017) proposed a model whereby PCD mechanisms were utilized to remove cells infected by viruses in the early evolution of animals, with the viruses acting as agents of natural selection for programmed cell death. The cells infected by viruses sacrifice themselves by PCD, killing the viruses that are infecting them, thus helping control the viral infection.

Segovia et al. (2003) showed PCD in a one-celled alga, suggesting PCD did not originally evolve in animals. From this work, they hypothesize that ancient viruses transferred critical DNA segments of the PCD pathway to the DNA of early unicellular eukaryotes in the Precambrian Ocean before the evolution of multicellular organisms, and they were then passed on to multicellular plant and animal lines as they evolved. The Precambrian supereon spans from about 4.6 bya to about 541 mya.

Thus, it appears that kin selection and viruses were both essential in the evolution of both prokaryotic and eukaryotic PCD, which is important in the evolution of multicellularity. Coevolution between viruses and their hosts was almost certainly one of the key factors behind the evolution of simple multicellularity, possibly on more than one occasion. This prepared the ground for the emergence of advanced multicellular organisms.

Viruses also played a major role in the evolution of advanced multicellularity in complex animals from one-celled eukaryotes. Viruses may have brought cells together to better spread to uninfected cells. Viruses transferred genes that are critical in the development and specialization of multicellular tissues and organs in higher animals. Two types of viral proteins that were important in the evolution of multicellularity have been identified thus far. First are syncytins, proteins discussed later in this chapter in the discussion on the origin of live birth in mammals. The other is EFF1, which helps form the skin of a species of round worm. Significantly for the evolution of multicellularity, EFF1 does the work of linking one cell to another in viral infections (Jamin et al., 2014). All known molecules that bring separate cells together came from viruses, suggesting viruses were crucial to the evolution of multicellularity and to the communication systems between cells that enabled multicellularity. *Genes in animals that came from viruses apparently give cells the ability to grow into tissues and organs, and even undergo sexual reproduction. Without these genes, animals could not have evolved beyond simple blobs of cells* (Slezak, 2014). Without the capacity of individual cells to join together into one organism, colonies of cells could have formed, but nothing even as complex as simple sponges could have existed. However, there are several other hypotheses and factors unrelated to viruses that might be important to the evolution of animals.

Viruses played key roles in major evolutionary breakthroughs in animals. *Proteins that are important in the regulating of animal development called hedgehog proteins were obtained from a peculiar class of parasites called inteins that combine jumping segments of DNA with shufflinWg parts of proteins around (called protein splicing), implying a viral or other parasitic, selfish DNA contribution to the origin of animals and their development* (Koonin, 2016).

205

Chapter Eleven

The gene called Arc is essential for long-term memory and learning in mammals.it helps with plasticity in synapses, the junctions between nerve cells. This gene codes for a protein that forms capsules that contain, protect, and transport RNA molecules between nerve cells. The RNA molecule is needed for nerve cell function. Arc is a repurposed, now inactive, jumping gene. Originally the protein that Arc codes for formed a capsule to protect and transport the genetic material of the virus. The viral protein capsule was converted to carrying Arc RNA, and became necessary for long-term memory and learning in the vertebrate brain about 375 mya (Day and Shepherd, 2015; Pastuzyn et al., 2018). Fish do not have Arc, but amphibians do. So, Arc first appeared in vertebrates in amphibians. Amphibians first appeared about when the Arc RNA first became useful in learning and memory. So here we have a virus that infected vertebrates and then was co-opted by vertebrates and put to use for learning and memory.

Endogenous retroviruses (ERVs) are viral elements and a type of jumping gene in the genome, embedded in the DNA, that resemble viruses and can be derived from them. *ERVs sometimes get packaged and moved within the genome with the result that they end up serving a vital role in gene expression or regulation* (Khodosevich et al., 2002; Kim et al., 2004). Apparently, some of them evolved from viruses, although some researchers think some viruses evolved from ERVs, since ERVs can mutate and become separate from their cell, or even pathogenic (Cotton, 2001). Many are remnants of viruses, consisting of parts of the virus from which they came. It is not known which genes and regions of the genome that regulate genes (turning them on and off) came from ERVs that were originally viruses as opposed to originating in their host genome. I am using my best judgement in listing those which I think originated from viruses. I may have erred in categorizing the ones I list here as of viral origin. However, even those ERVs that originated from their host's genome and not from viruses demonstrate the tendency of the genome, and hence life, to promote evolutionary breakthroughs and diversification, and thus the validity of the ABH.

Thousands of mammalian endogenous retroviral sequences have been repurposed by their hosts to create new host genes, and to regulate host

206

genes, substantially contributing to regulatory innovation in mammals (Fe-schotte and Gilbert, 2012).

ERVs are prominent in the genomes of vertebrates that have jaws. They are estimated to be 7 to 8% of the human genome (Lander et al., 2001; Belshaw et al., 2004; Nelson et al., 2004). ERVs may play a big role in many evolutionary events. The human genome project found several thousand ERVs in humans, which are organized into 24 families (Villarreal, 2001). Since ERVs are bad for and disruptive to the host genome and the host in the great majority of cases, mutations in host genes that inactivate or delete ERVs tend to be favored by natural selection. And ERVs and their hosts have undergone antagonistic coevolution, which is coevolution involving some conflict. This conflict may occur between competing species, a predator and its prey, a parasite and its host, or parasitic DNA and its host. This antagonistic coevolution resulted in an increase in the number of host genes that repress ERVs, stopping them from expressing themselves and making proteins. In some cases, these repressor genes have adapted to repressing host genes. If an organism's repressor genes turn off some of the organism's other genes, this can cause large, rapid, adaptive evolutionary change in higher organisms because repressing host genes can create new functions. For example, the gene could stop growth of a tissue, so turning it off would make the tissue grow larger, leading to a new, adaptive function. When birds evolved from dinosaurs, they evolved beaks by turning off the genes for making teeth in a big evolutionary breakthrough. The beak has many functions, including its use for building nests.

ERVs have been important in evolution, and are much used in the regulation and expression of genes (Khodosevich et al., 2002; Kim et al., 2004). ERVs are sometimes repurposed by the host to serve new functions for the host, especially in reproduction and development. They are often repurposed by the host to act as regulators of host genes, turning them on or off, often producing variants that regulate the development of specific tissues.

The gene AMY1C codes for an enzyme important in digestion in humans called amylase. Amylase is in the saliva of humans and some other mammals, where it begins the chemical process of digestion. It breaks down starch into sugars. The gene for it has a complete ERV sequence in its control region that turns it on and off, showing that this ERV was repurposed for regulation of amylase. A long terminal repeat is a pair of long identical repeating segments of DNA in eukaryotic genomes, each on one end of a jumping gene, EVR, or similar DNA entity. A long terminal repeat associated with this EVR regulates the expression of amylase in the salivary glands (Ting et al., 1992). The primary switch that turns on and off an enzyme that participates in the production of bile acid is from a long terminal repeat of an ERV (van de Lagemaat et al., 2003). Bile acids facilitate digestion of dietary fats and oils.

An upstream long terminal repeat of an EVR helps turn on a protein that humans need to utilize fats (Oliver and Greene, 2011). Note that the EVRs discussed here likely came from viruses that once infected humans and/or human ancestors, but which were repurposed by humans or human ancestors to take on the important functions listed here.

Long-terminal repeats are usually DNA segments that proteins bind with to turn genes on, with 64% of them expressed in reproductive organs (Romanish et al., 2007). There is a long terminal repeat involved in regulating estrogen production that is expressed in the brain and reproductive organs of most mammals; a variant of this in primates is responsible for controlling estrogen levels during pregnancy (van de Lagemaat et al., 2003).

In eukaryotes, DNA is in structures called chromosomes that also contain proteins that are bound up with the DNA. When sperm and eggs are made, the chromosomes line up and exchange their DNA. This is a source of variation which natural selection acts on, resulting in evolution. Similarly, ERVs and their long terminal repeats can trade pieces of DNA with each other and cause rearrangement of chromosomes. Sometimes when this happens, one EVR or long terminal repeat gets more DNA than it gives to the other. This can cause genes to double in number in the ones that receive more DNA than they give, and the loss of genes in the ones that give more DNA than they receive. This can sometimes change the function of genes (Hughes and Coffin, 2001). This can occasionally result in the appearance of beneficial changes (adaptations) to organisms. When there are duplicate copies of a gene, one copy can become a pseudogene and undergo maladaptive mutations for a long time while the normal gene covers for the pseudogene and carries out the function of the gene. This can happen until the pseudogene mutates to a lucky new adaptive function, causing an evolutionary breakthrough, as discussed under the section in this chapter on sexual reproduction.

There are many visible differences between humans and our very close relatives, chimpanzees. Obvious differences include greater amount of hair on the bodies of chimpanzees and the fact that humans walk on two legs, not four. Many of these differences are a result of differences in the regulation of genes shared by these species. These genes are basically the same in humans and chimpanzees, but they are regulated differently. Some are turned on in humans and turned off in chimpanzees, and vice versa. *Human-specific endogenous retroviruses and their solitary long terminal repeats are probable candidates for the regulators of these genes. Thus, human-specific endogenous retroviruses are likely responsible for many of the differences between humans and chimpanzees that are visible to us. So viruses were likely very important in the evolutionary divergence between humans and chimpanzees.*

There is clear evidence that components of viruses themselves actually became essential elements of the immunity that organisms use against viruses, that viruses provided crucial and coordinated features of the immune system in all organisms, from simple prokaryotes to complex eukaryotes (reviewed in Villarreal, 2011). This happened because the immune systems of their hosts received DNA from viruses.

In humans, the major histocompatibity complex (MHC) genes are very important to the immune system. They have a high number of ERVs (Gogvadze, et al., 2009). Sznarkowska et al. (2020) stated: "Viral-derived elements…are extremely dense in the MHC region, accounting for flexible expression of the … genes and adjusting the level of immune response to the environmental stimuli…. The complex regulatory network behind MHC expression is largely controlled by virus-derived elements." Thus, regulation of the MHC region is the result of the repurposing of viruses by their hosts. In other words, *a major part of our immune system*

is likely derived from viruses. Also in humans, human-specific ERVs provide protection against a wide range of antigens. An antigen is a toxin or
other unhealthy foreign substance which induces an immune response
in the body. Human-specific ERVs may have given humans an immune
system superior to other primates (Dawkins et al, 1999).

Viruses that are specific to each species make up a substantial proportion of all organisms' genomes, including the portion that regulates genes
(Sznarkowska et al., 2020, and references therein). *Endogenous retroviruses
and defective versions of them that were derived from them have been shown
to actively shape the structure and regulation of basically all species' genomes
(ibid.). This means the structure and regulation of the genomes of virtually all
species has been and still is actively shaped by viruses. Much of the regulation
of the genes of an organism is accomplished by elements derived from viruses!*

Viruses were key in the evolution of reproduction and live birth in mammals that bear live young. Not all mammals bear live young. Monotremes,
a mammal group that includes duck-billed platypuses, lay eggs. There is
evidence indicating that the ancestors of modern live-bearing mammals coevolved with a virus that suppressed the immune system of its host to better
survive. It is thought that the host then co-opted some of the genes of this
virus. These suppressor genes then suppressed the attack on the fetus by the
mother's immune system, which would otherwise have happened because
the fetus is a foreign substance to the mother. This improved the ability of
the fetus to survive in its mother (Li et al., 2012). During pregnancy in
live-bearing mammals, ERVs are activated and produced in high quantities
during the implantation of the embryo. They are hypothesized to carry out
the function of suppressing the mother's immune system, protecting the
embryo from the immune system of the mother. They keep the mother's
immune system from attacking the embryo or rejecting it and causing a
spontaneous abortion. The ERVs apparently can suppress the mother's immune system because they are derived from infectious viruses that evolved
the ability to suppress their host's immune system to defend themselves
from it.

ERVs play an important role in the development of the placenta of mammals. The placenta is crucial in keeping the fetus alive in live-bearing mammals. It provides oxygen and nutrients to the growing fetus and removes waste products from its blood. Most cells have just one nucleus. Remember the nucleus (plural is nuclei) is the organelle of the cell that contains the DNA. In mammals, proteins called syncytins are responsible for the formation and function of cells with many nuclei called syncytiotrophoblasts (Black et al., 2010), which maintain nutrient exchange between the mother and the fetus. And they separate the fetus from the mother's immune system (ibid.), protecting the fetus from it. Retroviral fusogenic envelope proteins cause cells to fuse together to form the syncytiotrophoblasts. They were originally used by viruses to enter cells to parasitize and infect them. It is probable that their ability to fuse cells together helped viruses get from one host genome to another to infect uninfected genomes when they reached a high population in the cell they were already infecting. At any rate, viruses and these fusogenic proteins that they make were co-opted by the mammals that they once infected, and the fusogenic proteins are now used by mammals to fuse cells together and form the cells with many nuclei that maintain the exchange of nutrients between the mother and her fetus, and keep the fetus separate from the mother's immune system in placental mammals. Thus, the mother's immune system does not attack the fetus. We can see that viruses played a key role in the evolution of live-bearing mammals and their placentas. The placenta is necessary for live birth. It protects the fetus from the mother's immune system, provides it with nutrients, and removes its waste only because it is a fused structure of one large cell with many nuclei. This structure could not have evolved in mammals without the help of viruses. The placenta evolved in many different mammal groups independently by symbiotic coevolution with viruses.

It is important to remember that *all of the ERVs discussed above were probably originally viruses that were repurposed by their host to provide an important adaptive function for the host.*

About 8% of the human genome originates from viruses, not from our vertebrate ancestors (Feschotte, 2010). These were viruses that attacked us from the outside. Another 40% is made up of repetitive segments of DNA that are also thought to have a viral origin. So about 48% of our genome

ultimately comes from viruses. For comparison, only about 2% of our genome codes for proteins. Thus, the human genome has up to 24 times as
much DNA that it acquired from viruses as DNA that codes for proteins!
Over 100,000 DNA sequences in the human genome are derived from
ERVs. *As many as 85 genes derived from viruses might be functioning in
the brain and during pregnancy in humans.* It is reasonable to assume that
similar percentages hold for other eukaryotes, and that all higher organisms
are chimeras and that a large part of them is virus.

Endogenous viral elements (EVEs) are DNA sequences that came from
viruses found in non-viral organisms. They can be whole viruses or parts
of them. They are remnants of ancient viruses, that integrated into the
DNA of their host. They account for about 10% of mammalian genomes,
represent a large source of both coding and noncoding DNA, many have
been co-opted by their hosts for various functions, the principal functions they now provide their hosts with are immunity and development,
and those that mitigated conflict between virus and host might have been
stepping stones toward co-option for host processes like host development
(Frank and Feschottte, 2017). DNA that was formerly used by viruses for
infection and replication was repurposed by the hosts for these functions.

Horizontal Gene Transfer Provides Variation for Evolution, Promotes Evolution, and Can Cause Large Evolutionary Transitions

Horizontal Gene Transfer (HZT) is another mechanism by which the
genome of organisms promotes genetic and phenotypic variability, biodiversity, and evolutionary innovation. This is the lateral movement of genetic material between organisms, which is distinct from the transmission
of DNA from parent to offspring, which is called vertical gene transfer. It
is an important factor in evolution. HGT can occur between organisms of
the same species, between species, and even between unrelated species in
different taxa. The genes of any taxon can jump to any other taxon, regardless of how unrelated they are. HGT is so common that some scientists
consider it to be the most serious hidden and underestimated hazard of
genetic engineering (Ho, 2007). Syvanen (1985) correctly said that HGT
was involved in shaping evolutionary history from the beginning.

HGT is common in prokaryotes, and a large percentage of the genes in their genomes were acquired by HGT between species. Most of the of increase in the number of protein families and acquisitions of new protein functions are due to HGT in prokaryotes (Grassi et al., 2012). Their great diversity, ability to adapt rapidly, and ability to occupy a great variety of niches and habitats are to a large extent due to HGT. This is significant because prokaryotes evolved the vast majority of life's biochemical functions. These include anoxygenic and *oxygenic* photosynthesis, respiration, nitrogen fixation (converting molecular nitrogen, which life cannot use, to a form life can utilize), and the ability to break down and digest a good deal of plant nutrients that animals cannot digest.

Photosynthesis uses energy from sunlight to produce the carbohydrate the photosynthetic organism needs for food. Raymond et al. (2002) looked at five unrelated taxa of bacteria that carry out photosynthesis from very different, unrelated evolutionary lines. They found a set of 188 genes inherited from common ancestors of the five groups. It appears the different species shared their genes by HGT. Photosynthesis evolved through the merging of separate evolutionary lines that brought together independently evolving chemical systems, through the giving and taking of blocks of genetic material between different bacterial species. Different pieces of the system evolved separately in different species, possibly at least in some cases even to serve purposes different from their current photosynthetic function. The genes were then brought together into one species, making a new combined system, by the transfer of DNA segments between species of bacteria. Bringing together of genes from different species into one species and the evolution of photosynthetic systems likely occurred several times. Not all photosynthesis produces oxygen. Only oxygenic photosynthesis does so. It is known in plants, but first appeared in bacteria called cyanobacteria. Oxygenic photosynthesis did not evolve in one species of cyanobacteria. There are two major systems of oxygenic photosynthesis, called Type I and Type II (abbreviated RCI and RCII). These two systems evolved separately in completely different evolutionary lines of bacteria. Organisms can photosynthesize with one photosystem or the other, but no organism performs oxygenic photosynthesis without both systems. Cyanobacteria are the only prokaryotes with both photosystems and the only prokaryotes that carry out oxygenic photosynthesis. The evidence supports the idea that RCI and RCII evolved independently in two separate lines of cyanobacteria, and the two systems were then brought together by HGT, creating oxygenic photosynthesis (Hohmann-Marriott and Blankenship, 2011). We do not know whether the bringing together of different genes to create oxygenic photosynthesis was facilitated by a virus (or viruses) in any of the steps along the way or not, although I think it is probable. Furthermore, oxygenic photosynthesis evolved by many genetic processes in addition to HGT. These include the duplication of genes, deletion of DNA, movement and relocation of genes, and coevolution of a photosynthetic bacterium inside a prokaryote that resulted in a eukaryote that could carry out photosynthesis (Green, 2007). The evolution of oxygenic photosynthesis, one of the most important processes on Earth, led to the increase in atmospheric oxygen content that provided the energy needed for the evolution of all higher life. Photosynthesis also provides the source of energy (carbohydrates), directly or indirectly, for the vast majority of the organisms on Earth. It is the basis of a great majority of Earth's food webs.

Jumping segments of DNA also played a key role in the evolution of nitrogen fixation. The filamentous cyanobacterium, *Microcoleus chthonoplastes*, a cosmopolitan prokaryote that often builds layered mats of microbes in a variety of different environments, acquired its cluster of nitrogen-fixing genes through HGT early in its evolution (Bolhuis, et al., 2010). This is tremendously significant in showing the importance of HGT in evolution and biodiversity, since nitrogen fixation is so important in helping support so many different ecosystems. HGT is thus a fundamental process in structuring ecosystems and greatly increasing biodiversity.

Blanc et al. (2010) showed that *Chlorella variabilis* NC64A, a onecelled photosynthetic green alga that lives symbiotically in *Paramecium bursaria* (a one-celled organism that swims in water), likely acquired its ability to produce cell walls made of chitin, a complex fibrous substance, from the capture of metabolic genes by HGT from fungi, prokaryotes, or viruses of algae.

Ferns thrived until flowering plants, some of which are tall trees in forests where ferns live, evolved and shaded them out, between 300 and 130 mya, denying them access to sunlight, and causing many fern species to go extinct. But the ferns that survived acquired a gene for a light-sensing protein called neochrome, which allows ferns to detect dim levels of both red and blue light, enabling them to grow toward any light that reaches them through the forest (Li et al., 2014). This likely enabled them to prosper on shady forest floors. Ferns acquired this gene from hornworts, which are simple, rootless plants that descended from some of the first land plants (ibid.). The reproductive cells of both ferns and hornworts congregate on moist areas on the forest floor, where they came into close contact, and gene transfer could readily occur. This gene transfer and adaptive response to low light led to a tremendous diversification of ferns about 100 mya. About 80% of living fern species are a result of this diversification.

Deuterostomes are higher animals typically characterized by their anus forming before their mouth during embryonic development. They consist of three phyla, or major groups: hemichordates (acorn worms); chordates, which include vertebrates such as reptiles and mammals; and echinoderms, which include sea stars and sea urchins. They have 30 genes that are not found in other animals, but are in marine algae and bacteria, where several of these genes are involved in modifying sugars on cell membranes (Simakov, 2015). In the filter-feeding ancestor of the three aforementioned phyla, these genes are thought to be associated with physiological, metabolic, and developmental adaptations. It is unlikely these genes were passed to animals vertically from algae and bacteria and then lost by intermediates between the algae/bacteria and deuterostomes. *They were very probably acquired by horizontal gene transfer from algae and bacteria to the deuterostome ancestor at least 570 mya. This raises the possibility that deuterostomes would not have been able to come into being without these HGTs from bacteria and algae. At minimum, deuterostomes would likely be substantially different without the gene transfers.*

Plant cell walls are enclosures around each plant cell. They protect the plant cell. To utilize this resource, bacteria secrete enzymes that break down plant cell walls. These enzymes can break down pectin, a unique fiber found in fruits and vegetables, and cellulose, an important structural component of the cell walls of green plants and many forms of algae. Eukaryotes generally cannot produce such enzymes, so cannot digest pectin or cellulose themselves, but rely on their bacterial symbionts in their digestive system to do it for them. Most animals do not have the enzymes or the appropriate microbes to digest pectin and plant cell walls, so cannot digest certain plants such as grasses. The microbes produce enzymes the host cannot. Shelomi et al. (2016) have found that genes that code for enzymes that digest pectin have moved horizontally to stick insects, resulting in the ability of these insects to produce their own enzymes that break down pectin. Stick insects are long, thin insects that camouflage themselves by looking like sticks or leaves. Their genomes contain multiple genes for enzymes that beak down and digest pectin, allowing them to digest a greater variety of plants than they previous could, without the help of microbes. This frees them from dependence on microbes for this function. And it increases their digestive efficiency because it is more efficient for them to digest pectin themselves, rather than rely on microbes to do it for them. These genes are found in gamma-proteobacteria, the most common bacteria type in the stick insect microbiome, and on leaves the insects eat. They thus obtained the genes either from their digestive tract bacteria or indirectly from such bacteria that deposited them on their leaf food. Tests show that, while some of the new genes that digest pectin kept their original function, others actually evolved new, yet unknown, functions. Analysis showed the gene jumps occurred between 110 and 60 mya, before stick insects rapidly diversified into the approximately 3,000 species living today. This indicates that the gene transfer that conferred the new ability to digest pectin was a key innovation causing or at least greatly aiding the great diversification of stick insects.

Kirsch et al. (2012, 2014) showed that a gene that encodes an enzyme that breaks down pectin was horizontally transferred from a sac fungus (phylum Ascomycota) to a common ancestor of the superfamily Chrysomeloidea, the leaf beetles and long-horn beetles, and the superfamily Curculionoidea, the weevils, about 200 mya. This was followed by independent duplications in these two lineages. These are all plant-eating insects. This HGT allowed these insects to digest pectin and obtain energy and nutrition from it. This gene diversified into a large family of genes that code for enzymes that break down pectin in these lineages of insects.

As much as 40% of the human genome has moved around in our DNA during human evolution (Goodsell, 2006).

Transposable Elements

One way that HGT occurs is by the transposable element (TE), which is a mobile DNA segment that can move to another part of the genome of the same organism, to another organism of the same species, or to another species. It can sometimes carry some extra DNA that is not normally part of it from the genome it is moving from with it. Movement of TEs is important in creating genetic variability within species, giving them the ability to adapt to changing environmental conditions (Reznikoff, 2003). It is another mechanism by which the genome promotes variability, adaptive evolution and macroevolution, and diversification. Movement of TEs is common in prokaryotes (Touchon and Rocha, 2007), and often serves as a mechanism to transfer genes between bacterial species (Frost et al., 1985). *TEs are very common, frequently occupying large portions of a given genome. They are the single most abundant entity of large eukaryotic genomes. They make up more than 85% of the maize genome* (Schnable et al., 2009), *almost 50% and at minimum 45% of that of humans* (Lander et al., 2001; Deininger and Batzer, 2002), *40% of the mouse genome* (Waterston and Pachter, 2002), *and almost half of that of the average mammal* (Percharde et al., 2018). *Genes coding for enzymes that help the movement of TEs to another part of the genome are quite widespread in the genomes of most species and are the most abundant genes known* (Aziz et al. 2010). *That enzymes that help TEs move evolved and are common is evidence for my hypothesis that there is selection for a favorable amount and type of variability of the genome.*

TEs provide a source of genetic variability on which selection can act. Their effect on evolution can be both immediate and lasting (for reviews, see Oliver and Greene, 2009; Zeb, et al., 2009; Cordaux and Batzer, 2009;

217

and Schaack et al., 2010). Their primary significance to adaptive evolution seems to be their ability to induce changes in the regulation of genes, or the coding potential of genes (see, for example, Gill et al., 2021), without destroying existing gene functions. Changes in the regulation of genes can cause large, rapid changes in organisms that preserve the compatibility of the different parts of the genome.

Koonin (2016) outlined how viruses and TEs played key roles in major transitions in evolution, including the origin of protocells, the prokaryotic cell, the eukaryotic cell, multicellularity, and cellular defense systems, accomplished by a combination of selective pressure on the hosts from parasitism by the viruses and TEs, cooperation between parasite and host, and by the donation of DNA and from virus and TE to host as well as the co-opting of viruses and TEs to adaptive functions by the hosts. The origin of the sophisticated type of sex done by eukaryotes at the cellular level, called meiosis, in which the chromosomes line up at the center of the cell and move apart, is a key part of the major transition that led to the origin of eukaryotes. Koonin (2016) also hypothesized that TEs were fundamental in the evolution of meiosis.

Berkemer and McGlynn (2020) analyzed thousands of family trees derived from the comparison of DNA similarity data from thousands of microorganisms. They found that before the three domains of life separated, evolution was much faster, mutation rates much higher, and there were many more TE's moving between organisms than today. TEs were thus very important in evolution during the beginning phase of life.

It is astonishing that nearly 50% of the human genome is made up of TEs acquired from outside of it (Lander et al., 2001; Cordaux and Batzer, 2009). It is possible that at least some of this DNA is from symbiotic microbes in the human microbiome via HGT to human DNA (see Dunning Hotopp et al., 2007). Eukaryotes are truly genetic chimeric mosaics of DNA combined from different types of organisms. It is true that much of this DNA is thought to be selfish, parasitic DNA with no function to the eukaryote, but some TEs may have been instrumental in creating new functions (Oliver and Greene, 2009).

Nekrutenko and Li (2001) found that TE integration into the DNA often has an effect on gene function and TE insertion might create new genes. They stated that there is increasing evidence for an important role of TEs in gene evolution.

Van de Lagemaat et al. (2003) found evidence that TEs affect the expression of many genes through the donation of regulatory signals. They found that classes of genes that changed a good deal and expanded in the recent past, such as those involved in immunity or response to external stimuli, have many TEs, while genes with basic functions in development or metabolism that did not change very much over time lack TEs. Their results support the view that TEs have played a significant role in the evolution and diversification of mammalian genes.

The Microbiome Can Supply Genes to its Eukaryotic Host

There is a tremendous diversity of viruses, bacteria, archaea, fungi, and invertebrates in and on all multicellular species. Any of these small organisms has DNA segments and genes with the potential to jump into the genome of its eukaryotic host. Most of these transfers of DNA from the so-called microbiome would be disadvantageous to the host. However, at times these DNA transfers could bring new genes and new functions, and provide adaptive changes and even large evolutionary breakthroughs to the eukaryotic host. Heheman et al. (2010) found genes for enzymes to digest a marine alga were transferred from a marine bacterium to a bacterium in the digestive tract of Japanese people, allowing them to digest the marine alga. Although the genes were not transferred to humans, the finding suggests the possibility.

Polyploidy Can Contribute to Large, Adaptive Evolutionary Revolutions

Although not the norm, the entire genome can duplicate in just one generation. This usually occurs because chromosomes do not separate during meiosis, the process of cell division that forms the sperm or egg.

This results in an organism with double the normal number of chromosomes that its species normally has. It may happen twice, resulting in four times the normal chromosome number, or any number of ways and times, resulting in three times the normal chromosome number, six times it, and so on. The state of having more than the normal chromosome number is called polyploidy, and polyploidy can also mean the process that produces a polyploid. The production of polyploids results in the production of a new species, since polyploids cannot breed with their ancestors. This is rapid evolution, for *it is speciation in just one generation.*

Autopolyploids are polyploids derived from one species. Allopolyploids are polyploids derived from two different species that hybridize with each other. The production of polyploids provides the raw material for great bursts of innovation. By the use of redundancy, it allows an entire set of genes to carry out normal genetic functions while another entire duplicate set of genes evolves as pseudogenes with many deleterious mutations until some of them hit on new, adaptive functions, producing new, useful genes. This is the same mechanism as that discussed in the section on sexual reproduction in this chapter. But in polyploidy, all of the genes in the entire genome are evolving in this way, so there are many possibilities of producing new, large, adaptive evolutionary breakthroughs. This means more than one gene could produce an adaptive evolutionary innovation, with many genes together causing large adaptive changes in the phenotype. Ohno (1970) suggested polyploidy played a major role in evolution.

Polyploidy increases diversity beyond just species number, since most polyploids display novel variation or differences in structure from their parental species. This may help them to use new niches that are different from the niches of their parental species. It could help them survive by reducing competition with their parental species. The differences polyploids have from their parental species can result from higher amounts of proteins produced because there are at least twice as many genes, the coming together of different gene regulation systems if the parental species are two different species, or rearrangements of chromosomes, all of which affect which genes are present and the regulation of these genes (Osborn, et al., 2003; Chen and Ni, 2006; Chen, 2007; Albertin, et al., 2006).

Polyploidy is common outside the animal and plant kingdoms. Some bacteria and archaea are polyploid. *Paramecium tetraurelia*, a one-celled eukaryote, duplicated its genome three successive times. Some diatoms, which are common one-celled algae in soil, freshwater, and the sea, apparently are polyploid. There is evidence for a polyploidy-producing event in the ancestry of yeast, and that this was followed by the loss of over 90% of the newly duplicated genes (Wolfe and Shields, 1997). Diverse species of fungus have had past and recent gene doublings, and this has often been followed by diversification in their phenotypes. Some brown algae seem to be polyploid.

Polyploidy is pervasive in plants, where it is estimated that *30–80% of living species are polyploid* (Meyers and Levin, 2006), and that 70% of all flowering plants have gone through one or more cycles of chromosome doubling. Wood et al. (2009) found that about 15% of the times flowering plants and 31% of the times ferns divide into new species, polyploidy occurs. They found that polyploids do not diversify into new species at a greater rate than plants with the original chromosome number, but after polyploid formation, considerable and sometimes very rapid changes in genome structure and the amount that different genes are expressed have occurred (Matzke et al., 1999). The coast redwood (*Sequoia sempervirens*) is a hexaploid (meaning it has six pairs of chromosomes), with 66 chromosomes (Xu et al., 20012). Aquatic plants include a large number of polyploids (Les & Philbrick, 1993). Evidence suggests that genome multiplication occurred more than one time in the evolution of the flowering plant, thale cress (*Arabidopsis thaliana*) (Blanc, 2000). Three duplications of the entire genome of this plant have been directly responsible for over 90% of its increase in chemicals that regulate its genes and in genes involved in its development in the last 350 million years (Maere et al., 2005). Additionally, this genome showed a great deal of rearrangement, with a patchwork of duplicated regions that indicated insertion of DNA segments, tandem duplication of DNA, inversion of DNA segments, exchange of DNA segments between different chromosomes, and deletion of DNA segments (Blanc, 2000).

Analyses of plant genomes whose DNA had been sequenced showed that *polyploidy was crucial in the evolution of seed plants* (Yuannian et al., 2011). A seed plant is any plant that produces seeds, and includes most of the familiar land plants, including all of the flowering plants and the conifers such as pines, but not ferns. This is a huge evolutionary leap; seed plants are a very important taxon. The ability to produce seeds allowed great diversification of seed plants.

Albert et al. (2013) sequenced the genome of a small tree with no common name, *Amborella trichopoda*, showing that the first flowering plant appeared as a result of a gymnosperm (this group includes conifers, such as pine trees, and is the ancestral group that gave rise to flowering plants) doubling its entire genome about 200 mya, and that the ancestral flowering plant was a polyploid with a large constellation of both novel and ancient genes that survived to play key roles in flowering plant biology. The additional genetic material gave the plants the potential to evolve new, never-before-seen structures, such as flowers. In fact, comprehensive analysis of plant genomes whose DNA had been sequenced showed that flowering plants evolved as a result of *two* genome duplications (Yuannian et al., 2011). Based on an analysis of the entire genome of the plant called thale cress, De Bodt et al. (2005) also state there is compelling evidence that flowering plants underwent two whole-genome duplications during their early evolutionary history. These gene duplications were crucial for the creation of many important developmental and regulatory genes found in living flowering plant genomes. Indeed, De Bodt et al. argue that these ancient polyploid events might have had an important role in the origin *and* diversification of flowering plants, today's dominant plants. Flowering plants are a major group of plants consisting of essentially all plants that produce animal-pollinated flowers, meaning that *polyploidy led to one of the biggest evolutionary adaptive breakthroughs in the history of life on Earth.* The evolution of flowers put plants in a new adaptive zone with numerous new niches available to them, allowing them to diversify into many new species. Flowering plants are in numerous symbiotic relationships, including with fungi at their roots, bacteria that supply useable nitrogen to them and live in their roots, animals that pollinate them, and animals that disperse their seeds. The evolution and diversification of flowering plants also led to the diversification into myriad new species of each in the groups that are symbiotic with them.

Polyploidy occurs in many diverse taxa of animals. A few hundred cases of polyploidy are known in reptiles, amphibians, fish, crustaceans, insects, and other invertebrates. Polyploidy is more common in invertebrates, such as flatworms, leeches, and brine shrimp, than in vertebrates (Otto and Whitton, 2000). In fish, there are polyploids in the salmon family (Salmonidae) and many in the carp and true minnow family (Cyprinidae) (Leggatt and Iwama, 2003). Some polyploid fish species have 400 chromosomes (ibid.). It is common in amphibians. The African clawed frog (*Xenopus laevis*) has several different species with as many as 12 sets of chromosomes (ibid.).

Soltis and Soltis (1999) stated that polyploidy has played a major role in the evolution of many eukaryotes, and that it can cause extensive and rapid genome restructuring. They say such changes can be mediated by transposable elements, and polyploidy could represent a period of transilience, during which genomic changes occur, potentially producing new gene complexes and promoting rapid evolution.

Creation of New Species and Evolution by Hybridization Between Two Different Species

Hybridization between two different species, called interspecific hybridization, can create new species and cause evolutionary innovations. This sometimes involves the formation of polyploid species, but this is not always the case. This mechanism is primarily the result of the behavior of organisms rather than the genome. Nevertheless, the genome plays a role. Speciation by interspecific hybridization is considered fairly common, if not widespread, in the plant world (Arnold, 1997). It is important in some plant groups (Linder and Reiseberg, 2004). Plants form polyploids more easily than animals, so have more polyploid species that are the result of interspecific hybridization. It is estimated that 2-4% of all flowering plants and 7% of all fern species resulted from polyploid hybridization between two different species (Otto and Witton, 2000). Speciation by hybridization is quite common among oaks, especially in the white oak group. There are even several hybrids between two oak species that differ from each other above the species level.

In animals, the lonicera fly (*Rhagoletis mendax* × *Rhagoletis zephyria*) is a species created by hybridization between the blueberry maggot (*Rhagoletis mendax*) and the snowberry fruit fly (*Rhagoletis zephyria*). Most ornithologists believe the bird called the great skua (*Stercorarius skua*) is a hybrid species between the pomarine skua (*Stercorarius pomarinus*) and one of the northern skua species (*Stercorarius* sp.) (Furness and Hamer, 2003). The Clymene dolphin (*Stenella clymene*) is a hybrid of the spinner dolphin (*Stenella longirostris*) and striped dolphin (*Stenella coeruleoalba*) (Amaral et al., 2014).

Interspecific hybridization occurs with higher probability when the habitat is changed or disrupted. The American toad (*Anaxyrus americanus*) and Fowler's toad (*Anaxyrus fowleri*) are found throughout eastern North America with their ranges overlapping throughout much of it. Disruption of habitat by humans likely caused hybridization between these two species in small isolated areas.

If there is a group of several rapidly diverging species, they can sometimes form multiple hybrid species, giving rise to a species complex. An example of this is several closely-related but morphologically different genera of cichlid fishes in Lake Malawi in Africa (Genner and Turner, 2011). Cichlid fishes are a highly diverse family of fishes that includes the important food fish, tilapia (which is actually the common name for many species), and popular aquarium fish such as angelfish (genus *Pterophyllum*). Experts think many of the species of the duck genus *Anas*, which contains mallards and teals, are hybrid species; many of the different species in this genus can breed with each other and produce fertile offspring.

Hybrid speciation without an increase in chromosome number created many sunflower species (Reiseberg et al., 2003). Reiseberg et al. (2003) compared the common sunflower (*Helianthus annuus*) and prairie sunflower (*H. petiolaris*), which grow in mild conditions of central and western North America, with three other sunflower species: the western sunflower (*H. anomalous*), desert sunflower (*H. deserticola*), and paradox sunflower (*H. paradoxus*), which have adapted to the extreme environments of the dry, sandy soils of Nevada and Utah and the West Texas salt marshes. The ones in the harsher environments were hybrids of the ones in mild ecologies. The hybrids were hardier than the parental species, with larger seeds, which reduced the chance of being buried by moving sand; more rapid root growth, helping them access water; and narrower and more succulent leaves, reducing water loss in the drier environment. They flowered more rapidly, which allowed them to take advantage of seasonal rain, and their roots took up salty minerals less, allowing better survival in salty soil. Reiseberg thinks the hybrids could have evolved into new species in about 50 to 60 generations, a very short time evolutionarily. This shows hybrid species and speciation by hybridization can cause the evolution of adaptive evolutionary innovations.

We can conclude that speciation by hybridization is common and often leads to evolutionary innovations. Cotton, coffee, tobacco, sugar cane, wheat, and maize are all hybrid species, whether through natural crossing or human breeding. The frequency of hybrid polyploidy in ferns could be as high as 90% (Soltis and Soltis, 1999). Hybridization is common and important in a great number of species, genera, and families (Arnold, 1997).

The transfer and sharing of genetic material between species and higher taxa, including all the way up to between domains, by mechanisms such as the incorporation of viruses into their hosts' genomes, transduction, HGT between different taxa, TEs, and hybridization between species, means the branching evolutionary tree commonly used today to illustrate the evolutionary relationships between species based on their shared ancestry is obsolete, and we need to replace it with a new, revised evolutionary tree. The current branching evolutionary tree, as illustrated in Figure 11.1(a), must be replaced by a branching-and-network tree with cross branches horizontally connecting different branches and taxa, as illustrated in Figure 11.1(b). Vertical transfer from one generation to the next as the only mechanism of transfer of

genetic material has been the dominant paradigm for a long time. We
need a paradigm shift from the vertical evolutionary tree to a mosaic
web, with cross branches that are horizontally linked and connected, in
addition to their standard vertical branching pattern.

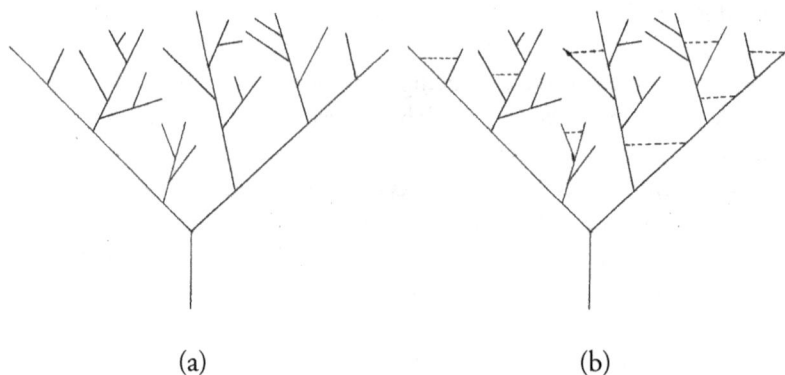

(a) (b)

Figure 11.1

(a). Evolutionary tree representing the old paradigm, which is a vertical
branching diagram representing the evolutionary relationships between
organisms or groups of organisms based on their shared ancestry. It
lacks horizontal cross branches connecting different evolutionary lines.

(b). Evolutionary tree representing the new paradigm, with horizontal
branches representing the horizontal transfer of genetic information
between evolutionary lineages. Horizontal lines are dashed instead
of solid to represent that there is only the transfer of DNA and not
horizontal speciation or evolution to new taxa between lineages.

Deletion of Segments of DNA

Deletion of segments of DNA is another mechanism for the evolution of
new genes and functions, and has occurred numerous times in the history
of the evolution of genes. It is another way the genome tends to increase
variability and biodiversity. Examples of deletions are found in the evolution of the genes for three of the four hemoglobin protein chains. Recall

that hemoglobin evolved by gene duplication. Deletions occurred in the evolution of each of the genes for the alpha, gamma, and delta hemoglobin chains.

A high prevalence of deletion mutations in a regulatory gene that turns other genes on has resulted in the loss of pelvic fins in some three-spined stickleback (*Gasterosteus aculeatus*) (fish) populations (Chan et al., 2010). The pelvic fins in fish are the paired fins at the bottom of the fish that are homologous with (match) the hind limbs of four-legged vertebrate animals such as horses.

Androgens are a group of hormones that play a role in male traits and reproductive activity. Receptors are molecules in cell membranes which respond specifically to a particular chemical, causing changes in the organism. Deletion of control genes that are enhancers for androgen receptors caused the human penis to lose spines, which are present on the penis of chimpanzees, other great apes, and monkeys (Reno et al., 2013; McLean et al., 2011 and references therein). These spines on the penis are sensory and produce pleasurable sensation and so cause ejaculation to occur more rapidly. Loss of spines is correlated with longer copulation time. This could have been coupled with tighter bonding during intercourse; long-term, faithful pair bonds; and monogamy in human ancestors. The male hunted and gathered food and brought it to the female and young in human ancestors. This is associated with the three traits just mentioned above. Fossil evidence suggests the reason that natural selection strongly favored walking upright on two legs in the evolution of the human line was that it freed the hands for the male to carry food to his mate and offspring (Lovejoy, 1981, 1988). Freeing the hands due to upright walking also allowed human ancestors to make and use tools and manipulate the environment, selecting for greater intelligence because brain power would have allowed more and better use of the hands for these functions. Cooperation and the providing of resources such as food by the male, evolving in conjunction with monogamous pair bonding between the male and female, could have lengthened the period of dependency of the offspring on the parents, making the time of learning as a juvenile longer. This would have made intelligence more favorable and adaptive, and caused natural selection to favor it. All of these traits could have evolved together. So, in humans, loss of spines on the penis due to gene deletions could have evolved together with a suite of traits, including walking upright on two legs, the providing of resources and food by the male, monogamy, pair bonding between males and females, male-to-male competition, extended learning periods of offspring, and intelligence. These would have led to macroevolutionary change and humans becoming highly intelligent. The deletion of the regulatory enhancer genes for androgen receptors also caused loss of facial vibrissae in humans; these are the "whiskers" apes and many other mammals have on their face and use to sense their environment. The deletion that caused these losses also occurred in the extinct human relatives, Neandertals and Denisovans.

McLean et al. (2011) confirmed 510 deletions of DNA segments in humans that are not coded and are enriched near genes involved in steroid hormone signaling and nerve function. Thus, it would seem they control genes for hormone signaling and nerve function. In chimpanzees and other mammals, these genes have not been deleted and are highly conserved. One deletion removes a regulatory gene enhancer for the human forebrain, and this deletion is correlated with expansion of specific brain regions in humans. That is, this deletion of a regulatory region that activates a gene that destroys excess primate neurons during embryonic development may help account for why human brains are much larger and have many more neurons than chimpanzees. The gene that destroys the neurons is not activated because part of the region that activates it was deleted. Deletion of a regulatory region that activates the gene involved in the growth of the bones of the toes helped us evolve our upright gate (Indjeian, et al., 2016). Upright walking is easier with shorter toes. *It is miraculous that deletions in the human genome are important in the intertwined evolution of intelligence, pair bonding, male provisioning, walking upright on two legs, and having hands free to use.*

Regulatory Genes and Evolution

Genes are arranged in a hierarchical system. Regulatory or control genes control the genes that code for RNA and proteins, switching them on and off. A mutation causing a small change in the DNA base sequence of a master control gene can have a very large effect on an organism's phenotype, including structure, color, and behavior. This is another mechanism by which a mutation, even one with a large effect, can occur and yet maintain the compatibility, coordination, and integration between the genes and different parts of the phenotype. Mutations in control genes can have large effects on the phenotype, and still maintain the compatibility of the many genes, and tissues, organs, and behaviors. The compatibility and integration of the genotype and phenotype have a high probability of being maintained in mutations of control genes because such mutations tend to cause the entire organism to change in a coordinated fashion, such as an increase in size by the same proportional amount in all parts of the organism. The mutation thus has an increased chance of being adaptive compared to mutations in structural genes (genes that do not regulate other genes). A large phenotypic change can occur in one generation. For a hypothetical example, a mouse could have a mutation in a regulatory gene, causing it to produce more growth hormone, which would cause the entire organism to increase in size, having all body parts enlarged equally without disruption of the system. This way, a mouse could hypothetically evolve into a rat.

The ancestor of vertebrates is a sea squirt, also called a tunicate, which is in the phylum that includes humans and fish, chordates, but a different subphylum, urochordates (human's subphylum is the vertebrates). The adult sea squirt is an invertebrate animal that cannot move and is attached to rocks or other hard surfaces on the seafloor as an adult. It is shaped like a flexible, vertical tube, and obtains its food by filtering small organisms out of the seawater. It has a larva called an ascidian tadpole that is shaped somewhat like a tiny eel, and has a head, nervous system, and tail, and can swim. The larva swims around until it finds a favorable place to settle on and change into the adult sea squirt on the seafloor. The evidence supports the hypothesis that in at least one ascidian tadpole, there was a mutation in a control gene that caused the larva to never change into an adult, but rather be able to breed as a free-swimming larva (Delsuc et al., 2006, and references therein). This is called *neoteny*, which is the sexual maturity of an animal while it is still in a larval state, or the retention of juvenile features in the adult animal. The neotenic, reproductive ascidian tadpole gave rise to the ancestor of fish and to the vertebrate line. This regulatory gene mutation achieved a huge maroevolutionary breakthrough in just one generation while conserving the integration and coordination of both the genotype and phenotype of the organism. All systems and parts remained integrated and compatible with each other, because the ascidian tadpole remained intact. The large macroevolutionary breakthrough allowed the reproductive ascidian tadpoles that were our vertebrate ancestors to enter into a new adaptive zone. As is the case with any macroevolutionary breakthrough leading to the entry into a new adaptive zone, it was followed by tremendous diversification into many new species, filling the many new niches made available as a result of being in a new adaptive zone. In this case, it was the tremendous number of species in the early vertebrate line. This shows how a mutation that causes a tiny change in the DNA in a control gene can rapidly cause a large, coordinated adaptive change and evolutionary breakthrough and transition, even in one generation. It is possible that a virus caused this change in the control gene.

In addition to a regulatory gene increasing or decreasing how much RNA or protein a structural gene produces, the output from a gene can increase by gene duplication. Both of these mechanisms can greatly alter the phenotype in an adaptive way without disrupting the integration or co-adapted nature of the genome and organism, leading to an evolutionary breakthrough.

Finally, a study of 12 animal species showed that after genes duplicate, they often change their expression pattern (Kryuchkova-Mostacci and Robinson-Rechavi, 2016). This can happen by the gene being expressed in different tissues and thereby adopting new roles. This would occur by changes in regulation of the gene.

There are numerous mechanisms of regulation of the genes, controlling the quantity of proteins each structural gene produces. Each is subject to mutation and selection, so can profoundly affect evolution. DNA codes for RNA in a process called transcription (the verb is to transcribe). The RNA then makes the protein in a process called translation. Protein is what DNA ultimately codes for, and genetic regulation is regulating the amount and timing of protein coded for by DNA. There are proteins called transcription factors which bind to specific DNA sequences on a gene, controlling when and how much of the gene is transcribed into RNA, affecting the amount of protein produced. They can activate or repress the gene. There are segments of the DNA involved in its regulation, such as promoters and operators, that proteins bind to for regulation of the genes. There are master regulatory genes that control many genes. Enzymes can add a methyl group, a small molecule of one carbon and three hydrogen atoms, to DNA, and this turns genes on and off. Enzymes can add an acetyl group, a small molecule of one oxygen, two carbon, and three hydrogen atoms, to proteins associated with DNA called histones, and this controls gene expression. Histones are involved in DNA structure and regulation. The tightness of winding of DNA around histones is involved in gene regulation, and this is controlled by chemicals in the cell. Tightly-packed DNA can prevent transcription factors from accessing genes. All of this is regulation of the genes at the level of transcription, direct control of the DNA and its activity. While most regulation occurs at this level, there is regulation of the RNA after transcription and before it codes for the protein, at the level of translation. Small RNA molecules called microRNAs decrease gene expression by inhibiting translation of RNA, degrading it, or by having silencer regions that bind to repressor proteins that inhibit the translation of RNA to protein. MicroRNAs also bind to microRNA response units that act as regulatory elements. There are 1,881 known microRNAs in humans alone. RNA can be regulated as to how much protein it manufactures by the binding of proteins and other RNA as well. Short interfering RNA is double-stranded RNA that interferes with the expression of specific genes by degrading coding RNA. RNA is also regulated by what is called anti-sense RNA and by long noncoding RNA, which is represented by an estimated 16,000 to 100,000 genes in humans. Some RNAs can self-regulate their expression. Thus, RNA molecules of a different type than RNA that codes for protein regulate this latter RNA. Even after the protein is made, it is regulated by chemical modification, how fast it is degraded, and other means. This is not a comprehensive list of the mechanisms of regulation of the genes, and each of these factors may involve many different molecules. Every one of these mechanisms is subject to mutation and selection. While almost all mutations will be deleterious, they have a higher probability of being adaptive because mutations of the system of genetic regulation have a better chance of maintaining the harmony of the genetic system and phenotype than random point mutations. The point here is that there are many ways that the system of regulation of the genes can mutate and produce a large, adaptive evolutionary transition in concert with natural selection.

Exaptation and Evolution

It is not uncommon for a gene, protein, or structure to change function in evolution, causing a change in the phenotype, sometimes a large one. An existing gene, protein, structure, or even behavior can be altered, causing it to take on a new function. In this process, called *exaptation*, selection acts on an existing structure of set of genes to create a novel function. Evolution acts by altering what is already present in the organism. If it occurs by switching the function of a gene, it is another example of how the nature of the genome can promote evolution, variability, and diversity. Exaptation can lead to new available niches, and thus diversification.

The FLO/LFY genes, called PpLFY1 and PpLFY2, regulate the first cell division and then the development of a moss called the spreading earthmoss (*Physcomitrium patens*). The same FLO/LFY genes underwent exaptation and in flowering plants are master regulatory genes for the development of flowers, without regulating cell division (Tanahashi, 2005).

Prolactin, a protein hormone, is thought to be related to regulation of water and salt balance in fish. It inhibits some species of salamander from changing from larvae into adults, causing them to breed as larvae and not undergo metamorphosis (Gona and Etkin, 1970). It promotes milk production in mammal mothers. It has evolved to take on several other functions in additional cases of exaptation. It is influential in more than 300 different processes in various vertebrates, including humans (Bole-Feysot, 1998). The hormone has important cell cycle-related functions in growth, differentiation, and countering programmed cell death. It is essential in metabolism, development of the pancreas, and regulation of the immune system. As a growth factor, it influences the formation of cellular components of the blood and the formation of new blood vessels. It is involved in the regulation of blood clotting. It has been modified to various other different forms in different taxa. All these functions of prolactin occurred by exaptation.

Microtubules are proteins that have been modified a number of times in evolution, taking on different functions by exaptation. There is a microtubule-like structure in the bacterium *Bacillus thuringiensis* involved in the separating of its small circular pieces of DNA called plasmids after they duplicate into two pieces (Jiang et al., 2016). The plasmids are separate from its main chromosome. Microtubules play a key structural role in the flagella and cilia of eukaryotes. Flagella are slender, microscopic, threadlike structures that enable many protozoa, bacteria, sperm, etc. to swim. They are like little whips attached to the cell. They generally attach to one-celled life forms, and there is generally only one or two per cell. Cilia are like little hairs on cells that are present in high numbers on each cell, and beat back and forth. They act as sensory organelles or to make the cell move. Microtubules form part of the cytoskeleton (this is like the skeleton of the cell) and provide structure and shape to eukaryotic cells. The microtubule cytoskeleton aids in the transport of material within cells. The cytoskeleton formed by microtubules is necessary to an organism's development; this has been shown in the fruit fly (van Eeden and St. Johnston, 1999) and mammals (Beddington and Robertson, 1999). The cellular cytoskeleton can also influence gene regulation (Rosette and Karin, 1995). Microtubules are essential in the development of the nervous system in higher vertebrates (Tucker, 1990). Microtubules act both to restrain cell movement and to establish its directionality in the developing organism. They play major roles in cell division, and are key in separating chromosomes during cell division in eukaryotes. In vertebrates, they are used in: the cilia of cells called epithelial cells that form the covering of internal and external surfaces of the body; the tail of sperm; and important parts of nerve cells. Finally, actin is a molecule that is important in muscle contraction. It is derived from microtubules by gene duplication. Thus, microtubules have undergone exaptation many times, taking on may new functions.

Phenotypic structures can undergo exaptation with great effects on evolution. Theropods are the group of dinosaurs that includes *Tyrannosaurus rex*. They were covered with feathers. These functioned to keep them warm and almost surely by males to attract females and intimidate rival males. When birds evolved from small theropods, feathers were exapted for use in flight. Arms of these small theropods also underwent exaptation at the same time that feathers did. They were originally used as arms are usually used. Then when they had feathers, they were used as nets to catch insects to eat. This helped them develop into wings. The wings were used for gliding before they were used for flight. The evolution of arms in dinosaurs to wings for flight in birds is a classic example of exaptation. The exaptation was accompanied by an evolutionary breakthrough and transition, with evolution into a new adaptive zone. So the birds diversified into many new species, taking advantage of the many new unoccupied niches available to them.

Experiments show that insect wings originally evolved and increased in size as a result of their ability to act as solar collectors to heat the insect up (Klingsolver and Koehl, 1985). They then took on the function of flight as insects increased in size and the wings increased in size proportionally to the insect's body.

Blankenship (1992) thinks that a good deal of the shared genes in photosynthesis were originally used by bacteria for purposes other than photosynthesis, and that these genes then changed functions to become part of photosynthesis systems by exaptation.

The air bladder of fish is used for buoyancy, allowing fish to swim at depths of their choosing. It evolved by exaptation into the lungs of amphibians when they evolved from fish. The fins of fish evolved into the legs of amphibians. Much later, legs used to walk on all fours eventually evolved into arms in up-right walking humans.

Exaptation is the rule. From the first prokaryote to eukaryotes, from simple to complex multicellular organisms, essentially every new gene is built from a pre-existing gene, and essentially every structure is built from a pre-existing structure.

Creativity of the Genetic Material Leads to an Enhanced Probability of Large Evolutionary Breakthroughs, Entry into New Adaptive Zones, and Hence Great Diversification into Many New Species

Most changes in the DNA are deleterious because the genome is a well-adapted, integrated system, and its compatibility will be disrupted by most changes. But there is a higher probability that the change will be adaptive if it is compatible with the genome and does not disrupt its existing compatibility. There is a higher probability of this if the change is the integration into the DNA of a segment of DNA that has undergone natural selection for generations in a genetic system that it evolved to be compatible with that does not differ significantly from the genetic system it enters, rather than a random mutation. The probability of an adaptive change is enhanced by a great number of trials. This number is high because of the high populations often present in evolving organisms and the great number of generations they evolve. If the genetic change is provided by a virus, the number of trials is enhanced by the number of viruses the host is exposed to. Viruses are the most numerous biological entities on Earth, outnumbering cells by a factor of 10 (Brüssow, 2009). The large number of ways large, compatible, adaptive change can occur (sexual recombination, viruses, TEs, control genes, polyploidy, etc.) also increases the chances of an adaptive change.

I propose that all of these mechanisms for rapid, large, adaptive evolutionary transitions collectively are the mechanism for the rapid change part of punctuated equilibrium, the idea that species do not change for long time periods, but occasionally undergo large, rapid changes (Gould and Eldredge, 1972). It is crucial to bear in mind that all of these mechanisms by which the genome is creative and has a tendency to promote adaptive evolution, instigate great diversification into many new species. This happens when any of the mechanisms discussed in this chapter cause a large, adaptive macroevolutionary change. All of the mechanisms are capable of this. The reason is that any large adaptive breakthrough or evolutionary transition, such as the evolution from fish to amphibians, puts the newly evolved taxon into a novel adaptive zone, which is a novel phenotype and lifestyle that gives the taxon access to many new ecological niches that are unoccupied. An example is when fish evolved into amphibians, amphibians had access to tremendous amounts of land, invertebrate prey, shelter provided by plants, all with no competition until they filled all the open niches. In addition, such a transition allows the new taxon to be free from predators for a good time period, also aiding diversification. Thus, the taxon diversifies to fill these niches. The result is diversification into many new species and a tremendous increase in biodiversity. This can be caused by integration of viruses into host DNA with the donation of genes to the host, TEs, changes in regulatory genes, deletions of DNA, or any other mechanism described in this chapter, and others that there was not enough space to discuss.

References

Watson, R. A. & Szathmáry, E. (2016). How can evolution learn? *Trends in Ecol. & Evol.* 31 (2): 147-57.

Aguilera, A. & Gomez-Gonzalez, B. (2008). Genome instability: A mechanistic view of its causes and consequences. *Nature Revs. Genet.* 9 (3): 204–17. doi: 10.1038/nrg2268

Kimura, M. (Feb., 1967). On the evolutionary adjustment of spontaneous mutation rates. *Genetics Research* 9 (1): 23-34. Pub. online by Cambridge University Press, 14 April 2009. doi: https://doi.org/10.1017/S0016672300010284

Lynch, M. (2010). Evolution of the mutation rate. *Trends Genet.* 26, 345–352.

Zhang, G. (2022). The mutation rate as an evolving trait. *Nat. Rev. Genet.* 24 (3).

Drake, J. W. (15 Aug. 1991). A constant rate of spontaneous mutation in DNA-based microbes. *PNAS USA* 88 (16): 7160-7164. https://doi.org/10.1073/pnas.88.16.7160

Dawson, K. J. (7 Sept. 1998). Evolutionarily stable mutation rates. *Journ. of Theoretical Biol.* 194 (1): 143-157. https://doi.org/10.1006/jtbi.1998.0752

Swings, T., Van den Bergh, B., Wuyts, S., et al. (2 May 2017). Adaptive tuning of mutation rates allows fast response to lethal stress in Escherichia coli. *eLife digest.* https://doi.org/10.7554/eLife.22939

Melamed, D., et al. (2022). De novo mutation rates at the single-mutation resolution in a human HBB gene-region associated with adaptation and genetic disease. *Genome Research* 32 (2). doi: 10.1101/gr.276103.121

Lander, E. S., et al. (International Human Genome Sequencing Consortium). (15 Feb. 2001). Initial sequencing and analysis of the human genome. *Nature* 409 (6822): 860–921. doi: 10.1038/35057062

Jeffreys, A. J., et al. (Oct. 2001). Intensely punctate meiotic recombination in the class II region of the major histocompatibility complex. *Nat. Genet.* 29 (2): 217–22. PMID 11586303. doi: 10.1038/ng1001-217

Baudat, F., et al. (2010). Prdm9 is a major determinant of meiotic recombination hotspots in humans and mice. *Science* 327 (5967): 836–40. doi: 10.1126/science.1183439

Myers, S., et al. (Aug. 2006). The distribution and causes of meiotic recombination in the human genome. *Biochem. Soc. Trans.* 34 (Pt. 4): 526–30. PMID 16856851. doi: 10.1042/BST0340526

Canchaya, C., Fournous, G., Chibani-Chennoufi, S., et al. (Aug. 2003). Phage as agents of lateral gene transfer. *Current Opinion in Microbiol.* 6 (4): 417-24. https://doi.org/10.1016/S1369-5274(03)00086-9

Anderson, N. G. (26 Sept. 1970). Evolutionary significance of virus infection. *Nature* 227: 1346 -7. doi: 10.1038/2271346a0

Liu, H., et al. (Nov. 2010). Widespread horizontal gene transfer from double-stranded RNA viruses to eukaryotic nuclear genomes. *Journ. Virol.* 84 (22): 11876-87. doi: 10.1128/JVI.00955-10

Claverie, J.-M. (2006). Viruses take center stage in cellular evolution. *Genome Biol.* 7 (6): 110. doi: 10.1186/gb-2006-7-6-110

Villarreal, L. P. & Witzany, G. (21 Feb. 2010). Viruses are essential agents within the roots and stem of the tree of life. *Journ. of Theoretical Biol.* 262 (4): 698-710. https://doi.org/10.1016/j.jtbi.2009.10.014

Filée, J., Forterre, P., & Laurent, J. (May 2003). The role played by viruses in the evolution of their hosts: a view based on informational protein phylogenies. *Research in Microbiol.* 154 (4): 237-43.

Villarreal, L. P. (30 March 2005). *Viruses and the evolution of life.* ASM Press: Washington, D. C. ISBN (Hardback): 1-55581-309-7. CABI Digital Library Record Number: 20053043558

Koonin, E. V. (19 Aug. 2016). Viruses and mobile elements as drivers of evolutionary transitions. *Philosoph. Trans. of the Royal Soc. B: Biol. Sciences* 371 (1701). This article is part of the themed issue, "The major synthetic evolutionary transitions". doi: 10.1098/rstb.2015.0442

Szathmáry, E. & Demeter, L. (21 Oct. 1987). Group selection of early replicators and the origin of life. *Journ. Theor. Biol.* 128: 463–86. doi: 10.1016/S0022-5193(87)80191-1

Szathmáry, E. & Maynard Smith, J. (1997). From replicators to reproducers: the first major transitions leading to life. *Journ. Theor. Biol.* 187: 555–71. doi: 10.1006/jtbi.1996.0389

Takeuchi, N., Hogeweg, P., & Koonin, E. V. (24 Mar. 2011). On the origin of DNA genomes: evolution of the division of labor between template and catalyst in model replicator systems. *PLOS Comput. Biol.* 7 (3): e1002024. doi: 10.1371/journal.pcbi.1002024

Higgs, P. G. & Lehman, N. (2015). The RNA world: molecular cooperation at the origins of life. *Nat. Rev. Genet.* 16: 7–17. doi: 10.1038/nrg3841

Shay, J. A., et al. (2015). The origin and spread of a cooperative replicase in a prebiotic chemical system. *Journ. Theor. Biol.* 364: 249–59. doi: 10.1016/j.jtbi.2014.09.019

Takahashi, I. & Marmur, J. (1963). Replacement of thymidylic acid by deoxyuridylic acid in the deoxyribonucleic acid of a transducing phage for Bacillus subtilis. *Nature* 197: 794-5.

Atkins, J. F., Gesteland, R. F., & Cech, T. R. (eds). (2010). *RNA Worlds: from Life's Origins to Diversity in Gene Regulation*, 3rd ed. Cold Spring Harbor Lab. Press., Cold Spring Harbor, N. Y.

Bernhardt, H. S. (2012). The RNA world hypothesis: the worst theory of the early evolution of life (except for all the others). *Biol. Direct.* 7: 23. doi: 10.1186/1745-6150-7-23

Gesteland, R.F., Cech, T., Atkins, J. F. (2006). *The RNA World.* Cold Spring Harbor Laboratory Press, Cold Spring Harbor, N. Y. 768 pp.

Forterre, P. (7 March 2006). Three RNA cells for ribosomal lineages and three DNA viruses to replicate their genomes: A hypothesis for the origin of cellular domain *PNAS USA* 103 (10): 3669-74. https://doi.org/10.1073/pnas.0510333103

Martin, W. & Koonin, E. V. (2006). Introns and the origin of nucleus-cytosol compartmentalization. *Nature* 440: 41-5.

Bell, P. J. L. (2001). Viral Eukaryogenesis: Was the ancestor of the nucleus a complex DNA virus? *Journ. of Mol. Evol.* 53: 251–6.

Bell, P. J. L. (7 Nov. 2006). Sex and the eukaryotic cell cycle is consistent with a viral ancestry for the eukaryotic nucleus. *Journ. of Theoretical Biol.* 243 (1): 54-63. https://doi.org/10.1016/j.jtbi.2006.05.015

Koonin, E. V. & Krupovic, M. (2020). Phages build anti-defence barriers. *Nature Microbiol.* 5: 8–9.

Chaikeeratisak, V., et al. (13 Jan. 2017). Assembly of a nucleus-like structure during viral replication in bacteria. *Science* 355 (6321): 194–7. doi: 10.1126/science.aal2130. PMC 6028185. PMID 28082593

Witzany, G. (2008). The viral origins of telomeres and telomerases and their important role in eukaryogenesis and genome maintenance. *Biosemiotics.* doi: 10.1007/s12304-008-9018-0

Schirrmeister, B. E. (2011). The origin of multicellularity in cyanobacteria. *BMC Evol. Biol.* 11: 45. doi: 10.1186/1471-2148-11-45

Iranzo J., et al. (1 Oct. 2014). Virus-host arms race at the joint origin of multicellularity and programmed cell death. *Cell Cycle* 13: 3083–8. doi: 10.4161/15384101.2014.949496

Quistad, S. D., et al. (2017). Viruses and the origin of microbiome selection and immunity. *The ISME Journ* 11 (4): 835–40. doi: 10.1038/ismej.2016.182. Pub. online 16 Dec. 2016.

Segovia, M., et al. (May 2003). Cell death in the unicellular chlorophyte Dunaliella tertiolecta. A hypothesis on the evolution of apoptosis in higher plants and metazoans. *Plant Physiol.* 132 (1): 99-105. doi: https://doi.org/10.1104/pp.102.017129

Jamin, M., et al. (10 April 2014). Structural basis of eukaryotic cell-cell fusion. *Cell* 157 (2): 407–19.

Slezak, M. (1 Mar. 2014). No Viruses? No skin or bones either. *New Scientist* 2958: 16.

Day, C. & Shepherd, J. D. (2015). Arc: building a bridge from viruses to memory. *Biochem. Journ.* 469 (1): e1–e3. https://doi.org/10.1042/BJ20150487

Pastuzyn, E. D., et al. (11 Jan. 2018). The neuronal gene Arc encodes a repurposed retrotransposon Gag protein that mediates intercellular RNA transfer. *Cell* 172 (1–2): 275-88. e18 https://doi.org/10.1016/j.cell.2017.12.024

Khodosevich, K., et al. (Oct. 2002). Endogenous retroviruses and human evolution. *Comparative and Functional Genomics* 3 (6): 494–8. doi: 10.1002/cfg.216

Kim, F. J., et al. (2004). Emergence of vertebrate retroviruses and envelope capture. *Virology* 318 (1): 183–91. doi: 10.1016/j.virol.2003.09.026

Cotton, J. (2001). Retroviruses from retrotransposons. *Genome Biol.* 2 (2): 6.

Belshaw R., et al. (April 2004). Long-term reinfection of the human genome by endogenous retroviruses. *PNAS USA* 101 (14): 4894–9. doi: 10.1073/pnas.0307800101

Feschotte, C. & Gilbert, C. (16 March 2012). Endogenous viruses: insights into viral evolution and impact on host biology. *Nature Reviews Genetics* 13: 283–96.

Nelson P. N., et al. (Oct. 2004). Human endogenous retroviruses: Transposable elements with potential? *Clinical and Experimental Immunol.* 138 (1): 1–9. doi: 10.1111/j.1365-2249.2004.02592.x

Villarreal, L. P. (Oct. 2001). Persisting viruses could play role in driving host evolution. *ASM News* (Amer. Soc. for Microbiol.) 67: 501–7.

Ting, C. N., et al. (1992). Endogenous retroviral sequences are required for tissue-specific expression of a human salivary amylase gene. *Genes & Dev.* 6 (8): 1457–65. doi: 10.1101/gad.6.8.1457

van de Lagemaat, N., et al. (Oct. 2003). Transposable elements in mammals promote regulatory variation and diversification of genes with specialized functions. *Trends Genet.* 19 (10): 530–6. doi: 10.1016/j.tig.2003.08.004

Oliver, K. R. & Greene, W. K. (2011). Mobile DNA and the TE-Thrust hypothesis: supporting evidence from the primates. *Mob. DNA* 2 (1): 8. doi: 10.1186/1759-8753-2-8

Romanish, M. T., et al. (2007). Repeated recruitment of LTR retrotransposons as promoters by the anti- apoptotic locus NAIP during mammalian evolution. *PLOS Genet.* 3 (1): e10. doi: 10.1371/journal.pgen.0030010

Hughes, J. & Coffin, J. M. (2001). Evidence for genomic rearrangements mediated by human endogenous retroviruses during primate evolution. *Nat. Genet.* 29 (4): 487–92. doi: 10.1038/ng775

Villarreal, L. P. (20 Oct. 2011). Viral ancestors of antiviral systems. *Viruses* 3 (10): 1933-58. https://doi.org/10.3390/v3101933

Gogvadze, E., et al. (2009). Human-specific modulation of transcriptional activity provided by endogenous retroviral insertions.

Journ. Virol. 83 (12): 6098–105. doi: 10.1128/JVI.00123-09

Sznarkowska, A. et al. (4 May 2020). MHC Class I regulation: The origin perspective. *Cancer* 12 (1155): 1-23. www.mdpi.com/ journal/cancers

Dawkins, R., et al. (1999). Genomics of the major histocompatibility complex: haplotypes, duplication, retroviruses and disease. *Immunol. Rev.* 167: 275–304. doi: 10.1111/j.1600-065X.1999.tb01399.x

Li, J., et al. (March 2012). Mouse endogenous retroviruses can trigger premature transcriptional termination at a distance. *Genome Res.* 22 (5): 870–84. doi: 10.1101/gr.130740.111. PMC 3337433. PMID 22367191

Black, S. G., et al. (2010). Endogenous retroviruses in trophoblast differentiation and placental development. *Amer. Journ. of Reproductive Immunol.* 64 (4): 255–64. doi: 10.1111/j.1600-0897.2010.00860.x. PMID 20528833

Frank, J. A. & Feschotte, C. (Aug. 2017). Co-option of endogenous viral sequences for host cell function. *Current Opinion in Virology* 25: 81-89. https://doi.org/10.1016/j.coviro.2017.07.021

Ho, M. W. (2007). *Genetic Engineering Dream of Nightmare? The Brave New World of Bad Science and Big Business.* Third World Network, Gateway Books, MacMillan, Continuum, Penang, Malaysia; Bath, UK; Dublin, Ireland; New York, USA. http://www.i-sis.org.uk/genet.php. ISBN-13: 978-9839747300. ISBN-10: 1858600510.

Syvanen, M. (Jan. 1985). Cross-species gene transfer; implications for a new theory of evolution. *Journ. Theor. Biol.* 112 (2): 333–43. doi: 10.1016/ S0022-5193(85)80291-5. PMID 2984477

Grassi, L., et al. (1 May 2012). Large-scale dynamics of horizontal transfers. *Mobile Genetic Elements* 2 (3): 163-7. https://doi.org/10.4161/mge.21112

Raymond, J., et al. (2002). Whole-genome analysis of photosynthetic prokaryotes. *Science* 298 (5598): 1616-20.

Hohmann-Marriott, M. F. & Blankenship, R. (2011). Evolution of photosynthesis. *Annu. Rev. Plant Biol.* 62:515-48.

Green, B. R. (2007). Evolution of light-harvesting antennas in an oxygen world. In Fawkowski, P. G. & Knoll, A.H., editors, *Evolution of Primary Producers in the Sea,* Ch. 4, pp. 37-53; Elsevier Inc., Amsterdam, Netherlands. https://doi.org/10.1016/B978-012370518-1/50005-9

Bolhuis H., et al. (Jan. 2010). Horizontal transfer of the nitrogen fixation gene cluster in the cyanobacterium Microcoleus chthonoplastes. *The ISME Journ.* 4 (1): 121-30.

Blanc, G., et al. (17 Sept. 2010). The Chlorella variabilis NC64A genome reveals adaptation to photosymbiosis, coevolution with viruses, and cryptic sex. *The Plant Cell.* doi: https://doi.org/10.1105/tpc.110.076406

Li, F., et al. (2014) Horizontal transfer of an adaptive chimeric photoreceptor from bryophytes to ferns. *PNAS USA* 111 (18): 6672-7.

Simakov, O., et al. (26 Nov. 2015). Hemichordate genomes and deuterostome origins. *Nature* 527: 459–65. doi: 10.1038/nature16150

Shelomi, M., Danchin, E. G. J., Heckel, D., Wipfler, B., Bradler, S., Zhou, X., & Pauchet, Y. (May 2016). Horizontal gene transfer of pectinases from bacteria preceded the diversification of stick and leaf insects. *Scientific Reports* 6: 26388. doi: 10.1038/srep26388. http://dx.doi.org/10.1038/srep26388

Kirsch, R., et al. (Nov. 2012). Combining proteomics and transcriptome sequencing to identify active plant-cell-wall-degrading enzymes in a leaf beetle. *BMC Genomics* 201213: 587. doi: 10.1186/1471-2164-13-587

Kirsch, R., et al. (Sept. 2014). Horizontal gene transfer and functional diversification of plant cell wall degrading polygalacturonases: Key events in the evolution of herbivory in beetles. *Insect Biochem. and Mol. Biol.* 52: 33-50. doi: 10.1016/j.ibmb.2014.06.008

Goodsell, D. (Dec. 2006). "Transposase". Molecule of the month. *Protein Data Bank.* doi: 10.2210/rcsb_pdb/mom_2006_12

Reznikoff, W. S. (March 2003). Tn5 as a model for understanding DNA transposition. *Molecular Microbiol.* 47 (5): 1199–206. doi: 10.1046/j.1365-2958.2003.03382.x. PMID 12603728

Touchon, M. & Rocha, E. P. C. (2007). Causes of insertion sequences abundance in prokaryotic genomes. *Mol. Biol. Evol.* 24: 969-81.

Frost, L. S., et al. (1985). Mobile genetic elements: the agents of open source evolution. *Nat. Rev. Microbiol.* 3: 722-32.

Schnable, P. S., et al. (2009). The B73 maize genome: complexity, diversity, and dynamics. *Science* 326: 1112-5.

Deininger, P. L. Mark A., & Batzer, M. A. (Nov. 2024). Mammalian retroelements. *Genome Research* 34 (11). http://www.genome.org/cgi/doi/10.1101/gr.282402

Waterston, R. H. & Pachter, L. Mouse genome Sequencing Consortium

(2002). Initial sequencing and comparative analysis of the mouse genome. *Nature* 420 (6915): 520–62. ISSN 0028-0836. https://resolver.caltech.edu/CaltechAUTHORS:20170309-090859678

Percharde, M., Lin, C.-J., Yin, Y., et al. (12 July 2018). A LINE1-nucleolin partnership regulates early development and ESC identity. *Cell* 174 (2): 391-405.e19

Aziz, R. K., Breitbart, M., Edwards, R. A. (1 July 2010). Transposases are the most abundant, most ubiquitous genes in nature. *Nucleic Acids Research* 38 (13): 4207–17. https://doi.org/10.1093/nar/gkq140

Oliver, K. R. & Greene, W. K. (2009). Transposable elements: powerful facilitators of evolution. *Bioessays* 31: 703–14.

Zeb, D. H., et al. (2009). Transposable elements and an epigenetic basis for punctuated equilibrium. *Bioessays* 31: 715-26.

Cordaux R. & Batzer, M. A. (2009). The impact of retrotransposons on human evolution. *Nature Reviews Genet.* 10: 691–703.

Schaack, S. et al. (Sept. 2010). Promiscuous DNA: horizontal transfer of transposable elements and why it matters for eukaryotic evolution. *Trends in Ecol. and Evol.* 25 (9): 537-46. https://doi.org/10.1016/j.tree.2010.06.001

Gill, R. A., Scossa, F., King, G. J., et al. (2021). On the role of transposable elements in the regulation of gene expression and subgenomic interactions in crop genomes. *Critical Reviews in Plant Sciences* 40 (2): 157–189. (Pub. online: 24 May 2021). https://doi.org/10.1080/07352689.2021.1920731

Berkemer, S. J. & McGlynn, S. E. (21 April 2020). A new analysis of archaea-bacteria domain separation: variable phylogenetic distance and the tempo of early evolution. *Mol. Biol. and Evol.* doi: 10.1093/molbev/msaa089

Dunning Hotopp, J. C., Clark, M. F., Oliveira, D. C. S. G., et al. (21 Sept 2007). Widespread lateral gene transfer from intracellular bacteria to multicellular eukaryotes. *Science* 317 (5845): 1753-1756. doi: 10.1126/science.1142490

Nekrutenko, A. & Li, W. H. (1 Nov. 2001). Transposable elements are found in a large number of human protein-coding genes. *Trends in Genet.* 17 (11): 619-21. https://doi.org/10.1016/S0168-9525(01)02445-3

Hehemann, J.-H., et al. (8 April 2010). Transfer of carbohydrate-active enzymes from marine bacteria to Japanese gut microbiota. Nature 464: 908–12.

Ohno, S. (1970). *Evolution by Gene Duplication*. Springer-Verlag, New York, N. Y.; Heidleberg, Germany; Berlin, Germany.

Osborn, T. C., et al. (2003). Understanding mechanisms of novel gene expression in polyploids. *Trends in Genet.* 19 (3): 141–7. doi: 10.1016/S0168-9525(03)00015-5. PMID 12615008

Chen, Z. J. & Ni, Z. (2006). Mechanisms of genomic rearrangements and gene expression changes in plant polyploids. *BioEssays* 28 (3): 240–52. doi: 10.1002/bies.20374. PMC 1986666. PMID 16479580

Chen, Z. J. (2007). Genetic and epigenetic mechanisms for gene expression and phenotypic variation in plant polyploids. *Ann. Rev. of Plant Biol.* 58: 377–406. doi: 10.1146/annurev.arplant.58.032806.103835. PMC 1949485 Freely accessible. PMID 17280525

Albertin, W., et al. (2006). Numerous and rapid nonstochastic modifications of gene products in newly synthesized Brassica napus allotetraploids. *Genet.* 173 (2): 1101–13. doi:10.1534/genetics.106.057554. PMC 1526534. PMID 16624896

Wolfe, K. H. & Shields, D. C. (1997). Molecular evidence for an ancient duplication of the entire yeast genome. *Nature* 387: 708–13.

Meyers, L. A. & Levin, D. A. (2006). On the abundance of polyploids in flowering plants. *Evolution* 60 (6): 1198-206. https://doi.org/10.1554/05-629.1

Wood, T. E., et al. (2009). The frequency of polyploid speciation in vascular plants. *PNAS USA* 106 (33): 13875–9. doi: 10.1073/pnas.0811575106. Bibcode: 2009PNAS.10613875W. JSTOR 40484335. PMC 2728988 Freely accessible. PMID 19667210

Matzke, M. A., et al. (25 Aug. 1999). Rapid structural and epigenetic changes in polyploid and aneuploid genomes. *BioEssays*. https://doi.org/10.1002/(SICI)1521-1878(199909)21:9<761::AID-BIES7>3.0.CO;2-C

Xu, X., et al. (2002). Origins of polyploidy in coast redwood (Sequoia sempervirens (D. DON) ENDL.) and relationship of coast redwood to other genera of Taxodiaceae. *Silvae Genetica* 51: 2–3.

Les, D.H. & Philbrick, C.T. (1993). Studies of hybridization and chromosome number variation in aquatic angiosperms: Evolutionary implications. *Aquatic Bot.* 44 (2–3): 181-228. doi: 10.1016/0304-3770(93)90071-4

Blanc, G., et al. (2000). Extensive duplication and reshuffling in the Arabidopsis genome. *Plant Cell* 12: 1093–102.

Maere, S., et al. (2005). Modeling gene and genome duplications in eukaryotes. *PNAS USA* 02 (15): 5454-9.

Yuannian, J., et al. (5 May 2011). Ancestral polyploidy in seed plants and angiosperms. *Nature* 473: 97–100. doi: 10.1038/nature09916

Albert, V. A., Barbazuk, W. B., dePamphilis, C. W., et al. (20 Dec. 2013). The Amborella genome and the evolution of flowering plants. Amborella Genome Project. *Science* (342 (6165). doi: 10.1126/science.1241089

De Bodt, S., Maere, S., & Vandepeer, Y. (Nov. 2005). Genome duplication and the origin of angiosperms. *Trends in Ecol. & Evol.* 20 (11): 591–7 doi: 10.1016/j.tree.2005.07.008. PMID 16701441

Otto, S. & Witton, P.J. (2000). Polyploid incidence and evolution. *Ann. Rev. of Genet.* 34: 401–37. doi: 10.1146/annurev.genet.34.1.401. PMID 11092833

Leggatt, R. A. & Iwama, G. K. (2003). Occurrence of polyploidy in the fishes. *Reviews in Fish Biol. and Fisheries* 13 (3): 237–46. doi: 10.1023/B:RFBF.0000033049.00668.fe. S2CID 11649126

Soltis, D. E. & Soltis, P. S. (1999). Polyploidy: recurrent formation and genome evolution. *Trends in Ecol. and Evol.* 14: 348-52.

Arnold, M. L. (1997). *Natural Hybridization and Evolution.* Oxford Univ. Press, New York, N. Y.

Linder, C.R. & Reiseberg, L.H. (22 June 2004). Reconstructing patterns of reticulate evolution in plants. *Amer. Journ. of Bot.* 91 (10): 1700–8. doi: 10.3732/ajb.91.10.1700

Furness, R.W. & Hamer, K. (2003). Christopher Perrins, ed. *Firefly Encyclopedia of Birds.* Firefly Books. pp. 270–3, Richmond Hill, ON, Canada. ISBN 1-55297-777-3.

Amaral, A. R., et al. (2014). Hybrid speciation in a marine mammal: the Clymene dolphin (Stenella clymene). *PloS ONE* 9 (1): e83645.

Genner, M.J. & Turner, G.F. (Dec. 2011). Ancient hybridization and phenotypic novelty within Lake Malawi's cichlid fish radiation. *Mol. Biol. and Evol.* 29 (Published online): 195–206. doi: 10.1093/molbev/msr183. PMID 22114359

Rieseberg, L. H., et al. (2003). Major ecological transitions in wild sunflowers facilitated by hybridization. *Science* 301 (5637): 1211-16.

Chan, Y. F., et al. (2010). Adaptive evolution of pelvic reduction in stickle-backs by recurrent deletion of a Pitx1 enhancer. *Science* 327 (5963): 302-5.

Reno, P. L., et al. (2013). A penile spine/vibrissa enhancer sequence is missing in modern and extinct humans but is retained in multiple primates with penile spines and sensory vibrissae. *PlOS ONE* 8 (12): e84258.

McLean, C. Y., et al. (2011). Human-specific loss of regulatory DNA and the evolution of human-specific traits. *Nature* 471 (7337): 216.

Lovejoy, C. O. (Jan. 1981). The origin of man. *Science* 211 (4480): 341-50. doi: 10.1126/science.211.4480.341

Lovejoy, C. O. (Nov. 1988). Evolution of human walking. Scientific Amer. 259 (5): 118-25. Secondary literature.

Indjeian, V. B., et al. (2016). Evolving new skeletal traits by cis-regulatory changes in bone morphogenetic proteins. *Cell* 164 (1): 45-56.

Delsuc, F., Brinkmann, H., Chourrout, D., et al. (2006). Tunicates and not cephalochordates are the closest living relatives of vertebrates. *Nature* 439: 965-8. https://doi.org/10.1038/nature04336

Kryuchkova-Mostacci, N. & Robinson-Rechavi, M. (Dec. 2016). Tissue-specificity of gene expression diverges slowly between orthologs, and rapidly between paralogs. *PLOS Computational Biology* 12 (12): e1005274. doi: 10.1371/journal.pcbi.1005274. PMC 5193323. PMID 28030541

Tanahashi, T., et al. (2005). Diversification of gene function: homologs of the floral regulator FLO/LFY control the first zygotic cell division in the moss Physcomitrella patens. *Development* 132: 1727-36. doi: 10.1242/dev.01709

Gona, A. G. & Etkin, W. (June 1970). Inhibition of metamorphosis in Ambystoma tigrinum by prolactin. *General and Comparative Endrocrinol.* 14 (3): 589-91. https://doi.org/10.1016/0016-6480(70)90042-0

Bole-Feysot C., et al. (June 1998). Prolactin (PRL) and its receptor: actions, signal transduction pathways and phenotypes observed in PRL receptor knockout mice. *Endocrine Reviews* 19 (3): 225–68. doi: 10.1210/edrv.19.3.0334. PMID 9626554

Jiang, S, et al. (March 2016). Novel actin filaments from Bacillus thuringiensis form nanotubules for plasmid DNA segregation. *PNAS USA* 113 (9): E1200-5. doi: 10.1073/pnas.1600129113. Bibcode:2016PNAS..113E1200J. PMC 4780641. PMID 26873105

van Eeden, F. & St. Johnston, D. (Aug. 1999). The polarisation of the anterior-posterior and dorsal-ventral axes during Drosophila oogenesis. *Current Opinion in Genetics & Development* 9 (4): 396–404. doi: 10.1016/S0959-437X(99)80060-4. PMID 10449356

Beddington, R. S. & Robertson, E. J. (Jan. 1999). Axis development and early asymmetry in mammals. *Cell* 96 (2): 195–209. doi: 10.1016/S0092-8674(00)80560-7. PMID 9988215. S2CID 16264083

Rosette, C. & Karin, M. (March 1995). Cytoskeletal control of gene expression: depolymerization of microtubules activates NF-kappa B. *The Journ. of Cell Biol.* 128 (6): 1111–9. doi: 10.1083/jcb.128.6.1111. PMC 2120413. PMID 7896875

Tucker, R. P. (1990). The roles of microtubule-associated proteins in brain morphogenesis: a review. *Brain Research. Brain Research Reviews* 15 (2): 101–20. doi: 10.1016/0165-0173(90)90013-E. PMID 2282447. S2CID 12641708

Kingsolver, J. G., Koehl, M. A. R. (1 May 1985). Aerodynamics, thermoregulation, and the evolution of insect wings: differential scaling and evolutionary change. *Evolution* 39 (3): 488–504. https://doi.org/10.1111/j.1558-5646.1985.tb00390.x

Blankenship, R.E. (1992). Origin and early evolution of photosynthesis. *Photosynth. Res.* 33, 91–111. https://doi.org/10.1007/BF00039173

Gould, S. J. & Eldredge, N. (1972). Punctuated equilibria: an alternative to phyletic gradualism. *Models in Paleobiology*, 82-115, ed. by Schopf, T. J. M. Freeman, Cooper & Co., San Francisco, CA.

Brüssow, H. (12 Aug. 2009). The not so universal tree of life or the place of viruses in the living world. *Philosophical Transactions of the Royal Soc. B. Biol. Sciences* 364 (1527). https://doi.org/10.1098/rstb.2009.0036

CHAPTER 12. IMPLICATIONS OF THE PACHAMAMA HYPOTHESIS FOR THE ENVIRONMENTAL CRISIS

Let us look at the implications to the environmental crisis of the Pachamama Hypothesis. All species make their environment better for life in a natural ecosystem, life made Earth better for life, and symbiosis and commensalism are of fundamental importance. What does this mean for people today, the current destruction of the Earth by humans, and what we should do? This will be a brief discussion and I will not attempt to make it a comprehensive one. I encourage the reader to think carefully about this discussion and come up with his or her own additional recommendations and conclusions.

If ecosystems maximize biodiversity and species promote biodiversity by enhancing their environment, this indicates biodiversity is favored by nature and is of value. It is inherently a good, beneficial asset that we should greatly value. protect, and conserve.

Every species in a natural system makes its environment better for life. This is partly the result of the fact that all species are profoundly interrelated, interconnected, and interdependent. Symbiosis and commensalism are examples of the interdependence of species. If any species, especially a keystone species, is removed from an ecosystem, other species decrease in number and may be extirpated from the ecosystem, meaning they will be driven locally extinct from the ecosystem, although not necessarily globally extinct. This means every species is important and worthy of preservation because at least some other species depend on it. A loss of any species likely means the loss of a number of species. Species are not separable from their habitat, on which they depend. One cannot preserve a species without saving its habitat. So we must preserve the habitats of species we wish to conserve. The conclusion to be drawn is that we must preserve not merely individual species or habitat fragments, but entire ecosystems because ecosystems are interconnected systems of species, and if any part of an ecosystem is destroyed, at least some other parts of the ecosystem will also be destroyed. Small species that most people do not notice or care about are especially worthy of protection. Often, these species are crucial to the health of their ecosystem. Mosquitoes pollinate flowering plants and are a source of food for lizards and birds. Soil bacteria keep the soil fertile and productive, and store carbon, keeping greenhouse gasses out of the atmosphere.

A corollary of the Pachamama Hypothesis is that a great many of our problems are alleviated by preserving biodiversity and made worse by destroying it. Forests, kelp, and other vegetation carry out photosynthesis, alleviating global warming and producing oxygen. Saving biodiversity preserves our medicines. Half of our medicines are derived from rainforests. This includes cures for some cancers, heart disease, fungal diseases, bacterial diseases, brain diseases, and malaria. Forests regulate the global and local water cycles and protect rivers and soil. Trees produce local rainfall and clean the air. Soil is built by organisms and held in place by trees and other plants. Rainforests regulate Earth's reflectivity, which affects rainfall patterns far from rainforests. Thus, destruction of the Amazon disrupts rainfall patterns in the US. Our food, various drinks (such as coffee), and spices come from animals, plants, fungi, and other organisms. Many of our products come straight from nature, from biodiversity. These include biodegradable, natural, non-toxic pesticides; houseplants; gums and resins; oils like coconut oil (used in baked goods, lotion, and soap); many kinds of wood; and canes and fibers. Wild yams from rainforests in Mexico and Guatemala give us diosgenin and cortisone, the active ingredients in birth control pills. Rainforests are also habitat to many indigenous people. Therefore, saving biodiversity would help with preserving global climate, protecting local climate, conserving soil, food production, production of many products we use, controlling many diseases, population regulation, air pollution, oxygen production, and many other problems I did not have space to mention here. I leave it as an exercise for the reader to think of more.

In addition, what we need to do to save biodiversity helps solve major problems. To preserve biodiversity, we need to burn a smaller quantity of fossil fuels to reduce climate change. We need to conserve our top soil, because species from soil bacteria to plants to soil animals are dependent on soil. We need to reduce noise, plastic, and chemical pollution to save the great number of species that are being eliminated due to these problems. Noise pollution in the sea is negatively impacting whales, and plastic pollution is having disastrous effects on a great deal of sea life. Acting positively on these problems also helps preserve biodiversity. There are many more examples of serious problems that would help preserve biodiversity if we made progress on solving them. Now, here are some specific recommendations.

We need to design cities with nature and biodiversity in mind. The term eco-city was coined by Richard Register (1987) in his book, in which he advocated for "rebuilding cities in balance with nature". Eco-cities are often described as meeting the following criteria: (1) they operate on a self-contained economy with locally-obtained resources; (2) they restore environmentally damaged urban areas; (3) they are well-planned to favor walking, biking, and public transportation over cars; (4) they maximize water efficiency and energy efficiency; (5) they promote recycling and reuse to create a zero-waste system; (6) they use and produce renewable energy and are fully carbon-neutral; (7) they produce their food locally, in or near the city; (8) they provide affordable housing and jobs for disadvantaged groups; and (9) most importantly for this discussion, they promote high biodiversity and nature in the city.

China has the world's largest eco-city development program, with hundreds of eco-city projects currently in development. The Sino-Singapore Tianjin Eco-City, a planned city jointly developed by the governments of China and Singapore located in Binhai, China, had about 100,000 residents, as of April, 2019. It has guidelines for wetland and shoreline protection, urban greenspace, air and water quality, water consumption, noise pollution, modes of transportation, waste production, local employment options, and more. Curitiba, Brazil is an eco-city that is a pioneer in sustainable urban planning, with many urban green spaces and an efficient bus rapid transit system.

We need cities that emphasize human needs and nature over the needs of the automobile. We need more native trees and other plants in our cities. These would bring down greenhouse gas levels and attract mammals, birds, reptiles, insects from beetles to butterflies, and other animal groups. Restoring and protecting creeks and nurturing native vegetation on their banks would help establish and rejuvenate amphibians, fish, crayfish, and various insect larvae.

We should build extremely compact cities with large buildings housing many people. Houses, places of work, schools, and shopping should all be in walking distance. This would make cars unnecessary, and save people money and time. People would be much healthier because they would get exercise from walking, and the air and water would be much cleaner. The compactness would mean that there would be a great deal of extra space for nature to exist. There would be much more land for natural ecosystems and biodiversity to thrive in a magnificent beauty of fluttering birds, butterflies, and dragonflies; colorful salamanders; acrobatic squirrels, and more.

In an amazing book, Camu (2024) advocates that people eliminate their ecologically-disastrous lawns and replace them natural ecosystems of native meadows and trees in their yards. If entire neighborhoods did this, we would see continuous forests and natural ecosystems of much higher biodiversity then we see in the typical grassy yard, with a greater variety of mammals, birds, reptiles, insects, and other invertebrates. Colorful butterflies would decorate our yards. Psychological health would increase as people became surrounded by a beauty they never experienced before in their yards and entire neighborhoods, and people became closer to and more in tune with nature. These actions would also draw down greenhouse gasses, which would create a more life-favorable temperature, improving people's health and further increasing biodiversity. These are practical actions the reader can do, improving his or her psychological and physical health and life satisfaction, while enjoying the beauty he or she created.

Our system of food production could benefit immensely by switching to a system that promotes sustainability and higher biodiversity than current agricultural practices. Most agriculture uses monoculture, the growing of but one crop, the lowest diversity possible. Monocultures are contrary to the tendency of ecosystems to maximize biodiversity. Polyculture is the practice of growing more than one species of crop or animal in an area at the same time, mimicking natural ecosystems. This increases yields, certainly increases biodiversity, improves soil health, and reduces the need for pesticides and fertilizers. It results in more varied habitats, promoting beneficial insects. Indigenous peoples of Central and North America plant beans, corn, and squashes together. These three crops work together in symbiosis, each species helping the others. The corn provides a three-dimensional support for the beans to climb and grow, reaching higher up toward the Sun for photosynthesis. Nitrogen is essential to all life, being used in DNA, RNA, proteins, and other molecules. But nitrogen in the air cannot be used by life. It must be converted to a form life can use, such as ammonia, in a process called nitrogen fixation. This is done mostly by nitrogen-fixing bacteria, some of which live freely in the soil and some of which live in the nodules, which are chambers, in the roots of certain kinds of plants. This is another example of the Pachamama Hypothesis: bacteria improving their environment for so many other species. Beans have the symbiotic nitrogen-fixing bacteria in their root nodules, and this symbiosis provides useable nitrogen in the soil to the beans, corn, and squash. The squash has fuzzy, prickly leaves that deter pests of all three species, and its broad leaves provide a natural mulch to suppress competing plant species and conserve soil moisture. Thus, increasing biodiversity creates a more stable system that increases yields and provides a sustainable, nutritious food source.

Indigenous peoples of Asia grow rice and fish together in one system. Fish feces fertilize the rice crop, enriching the soil with essential nutrients such as nitrogen and phosphorus. The fish eat insects and their larvae that consume the rice. They also consume plants that compete with the rice. As a side benefit, they eat mosquito larvae. The natural control by the fish reduces or even eliminates the need to apply toxic chemical pesticides and herbicides, which poison wildlife and native plants. This further protects diversity and results in a healthier ecosystem and safer, more natural food. Methane is a greenhouse gas that is 80 times as powerful as carbon dioxide, molecule for molecule. Soil microbes produce it in conventional rice fields, which can be big methane emitters, exacerbating global heating. Fish and their foraging can reduce methane emissions by up to 40%. The fish, along with other aquatic life in the water, stir up the soil, increasing its aeration and microbial activity, improving soil quality, water quality, and nutrient uptake by the rice plants. This also results in more efficient usage of the water.

More diverse systems are more stable. Even this small increase in diversity increases stability and resilience of the ecosystem. In turn, increased stability and resilience increase diversity, in a positive feedback loop. The increase in diversity is manifest in more species of beneficial plankton, algae, and insects. The fish do not consume the rice, but benefit because the insects, insect larvae, and plants that occur in this agroecosystem provide them with a food source.

Coffee benefits by being grown with trees that provide them with shade, mimicking coffee's natural habitat, where it thrives as an understory shrub in tropical forests. This is beneficial compared to full-sun cultivation in the following ways. It preserves the rainforest, a highly biodiverse ecosystem. Thus, it preserves biodiversity, in contrast to sun-grown monocultures, which have extremely low biodiversity. This preserves more birds, bats, and insects that are natural predators for coffee pests such as the coffee berry borer. This reduces or eliminates the need for chemical pesticides, which further protects wildlife and biodiversity, including fish and aquatic invertebrates because it prevents runoff of poisons into rivers. It also protects humans from cancer and other effects of pesticides. The tree canopy stabilizes the temperature, shields coffee plants from intense sun and rain and extreme weather, and increases humidity. This makes the farm more resilient to climate change effects like heat waves, frost, and drought. The shade trees absorb carbon dioxide from the atmosphere, storing up to 80 tons of carbon per hectare, compared to only 10 tons for sun-grown coffee. Agroforestry improves soil health because fallen leaves and branches, as well as dead trees, replenish the soil with nutrients. The deep root systems of trees also help prevent topsoil erosion and improve water retention of the soil. The shade slows down the ripening process of the coffee cherries, allowing more time for sugars and complex flavor compounds to develop, resulting in a more nuanced, sweeter cup with notes of chocolate, fruit, or flowers. This also creates a less acidic, more balanced, and smoother cup of coffee.

In Nigeria, as many as 13 crops are used in polyculture. These include, cassava, squash, pigeon pea, cowpea, maize, peanut, pumpkin, melon, and various yam species. This results in the same benefits listed above. In Indonesia, rice, fish, ducks, and water fern are grown together productively. The ducks eat the plants that compete with the rice, reducing the need for herbicides; the fish and duck feces fertilize the system; and the water fern fixes nitrogen.

Polycultures are more resilient and can reduce pests and diseases. Since different plants are susceptible to different diseases, if a disease attacks one crop, it will not necessarily spread to another. Hence, plant communities with higher biodiversity are less susceptible to diseases.

In polyculture, the combination of different plant species allows better use of light, water, and nutrients, and greater yield. Also, polyculture increases the biodiversity of the surrounding natural ecosystem. One of the mechanisms by which this is accomplished is the greater number of plant species attracts more pollinator species. These pollinate plant species in the nearby ecosystem.

Renewable energy, such as solar and wind, are necessary to conserve biodiversity because they produce less greenhouse gasses, and climate change is a major threat to biodiversity. But solar and wind have negative impacts on plants and animals. These can be mitigated to a certain extent, and positive environmental effects can be incorporated in their design. A few ways to do this follow.

Solar energy plants should be built on already-disturbed land (habitat) as much as is practical. Native plant species should be protected in the area of the plant, under and around solar panels. This will support pollinators and local wildlife. In hot areas, such as deserts, where the shade provided by solar panels is helpful to animals, the panels should be placed so animals can benefit from this. Fencing can block migration corridors for animals such as pronghorn, mule deer, and desert tortoises, and should be used as little as possible, and made to allow wildlife to pass through it where practical. In some cases, solar farms can act as wildlife corridors, connecting habitats. Reflective photovoltaic panels look like bodies of water to some birds, which at times attempt to land on them, leading to fatal collisions. So we should design these panels with anti-reflective coatings or patterned designs to reduce this risk.

We should mount panels high off the ground and increase spacing between arrays to allow native plants to grow underneath and create shelter for small animals. Large solar installations can increase local temperatures, and this must be mitigated, because it negatively impacts the plants and animals of the area. We should monitor the effects on wildlife and adapt our behavior according to what we find. And of course placing rooftop solar panels on homes and businesses and places like parking lots, and having community solar projects, allows solar energy generation without any habitat loss.

I will only discuss two examples of conserving diversity with wind farms. Radio-tagged individuals of the endangered California condor can be detected by wind farms, which causes their turbines in their path to stop turning turbines, preventing the birds from crashing into them and being hurt. The Fryslan Wind Farm in the Netherlands was built with radar monitoring and man-made islands to provide safe habitats for seabirds, demonstrating a case where wind farms enhanced local biodiversity.

Our transportation system depends mostly on cars and planes, both of which are major contributors of greenhouse gasses to the atmosphere. More than 350 million vertebrate animals are killed by cars in the U.S. each year, including hundreds of millions of birds and about 40 million squirrels. Road construction destroys and fragments habitats. Roads block animal movement and gene flow. Roads allow invasive species to enter native habitats. They cause predation rates to be higher than natural along forest edges, and impede animals from finding food, water, and mates. Traffic negatively affects animals psychologically with noise and light pollution. Traffic noise has caused birds to abandon their nests. Animals living near busy roads have higher levels of stress hormones. Hatchling sea turtles can be confused by road lights, causing them to walk away from the sea, which generally causes them to die. Rain and snowmelt carry pollutants from cars into waterways. These include chemicals that are toxic to aquatic life, such as copper, zinc, and lead from brake pads and tires, and oil and gasoline. The tire chemical 6PPD-quinone causes mass mortality in coho salmon. Mortensen et al. (2009)

found that non-native plants are twice as common with 500 ft. of a road as they are farther away. They found spread rates are higher in roadsides than in forested and wetland patches, even in the absence of major disturbances. Thus, road-building increases introduced species, which generally compete with and are bad for native plants. Roads in forests actually increase the frequency of forest fires.

Because planes have to overcome gravity to reach high elevations and move very fast, they burn great quantities of fuel and produce large quantities of greenhouse gasses. The FAA reported over 19,400 birds being struck by planes in the U.S. in 2023. Noise from planes leads to stress, psychologically-distressed behavior, and lower reproductive rates in mammals and birds. The large amount of air pollution from airplanes poisons wildlife and causes organ damage. Airport construction and expansion has destroyed a great deal of natural habitat.

The more we switch from cars and planes to public transportation, such as trains and busses, the more we will protect the climate and biodiversity. Many more people can be transported per passenger-mile by trains and busses than by cars.

References

Register, R. (1987). *Ecocity Berkeley: Building Cities for a Healthy Future.* North Atlantic Books, Berkeley, CA. ISBN: 9781556430091, 1556430094

Camu, B. (2024). *From Wasteland to Wonder: Easy Ways We Can Help Heal Earth in the Sub/Urban Landscape.* Leaf and Limb, Raleigh, NC. ISBN: 979-8-9900303-0-5 (print); 979-8-9900303-2-9 (digital online).

Mortensen, D. A., Rauschert, E. S. J., Nord, A. N., & Jones, B. P. (2009). Forest roads facilitate the spread of invasive plants. *Invasive Plant Science and Management* 2:191–199. https://doi.org/10.1614/IPSM-08-125.

GLOSSARY

adaptive zone   A broad area of adaptation. A set of related ecological niches. When a species invades a new habitat or evolves a new function, it typically enters a new adaptive zone. This allows it to diversify into many species that then occupy the many new, unoccupied niches available to it. An example of an adaptive zone is all niches for day time vertebrate flight. When birds evolved, they moved into this adaptive zone and diversified into the many niches in this zone, from niches of hummingbirds to eagles. Some other authors define this term somewhat differently than the way it is used and defined in this book.

anoxygenic photosynthesis   A chemical reaction that uses the energy of sunlight to convert hydrogen sulfide and carbon dioxide to carbohydrate (food) and elemental sulfur. It is a form of photosynthesis that does not produce oxygen.

antagonistic coeveolution   Coevolution between two species involving some conflict, such as competitors, herbivore and plant, predator and prey, and parasite and host. For example, a plant evolves a poison to guard against a herbivoire, the herbivoir evolves a defense against the poison, the plant evolves a stronger poison, and so on.

archaea (singular, **archeon**)   Prokaryotes that constitute a domain of life, and are not bacteria. The other two domains of life are bacteria and eukaryotes. Some archaea live in extreme habitats, such as hot springs with very high temperatures.

Autocatalytic Biodiversity Hypothesis (abbreviated **ABH**; also called the **Pachamama Hypothesis**)   The hypothesis of the author of this book, David Seaborg, that this book presents evidence and arguments for. It states that all species are ecosystem engineers that have a net positive effect on other species, their ecosystem, and biodiversity, in natural ecosystems, over sufficiently long time periods to make a fair assessment of their effects. It also postulates that ecosystems maximize biodiversity.

It posits that both symbiosis and commensalism are of fundamental importance in structuring ecosystems and increasing biodiversity. It has similarities to the Gaia Hypothesis and was inspired by it, but is distinctly different from it. It is original, profound, radical, supported by the evidence, scientific, and testable. It does not propose New Age ideas such as the Earth is a conscious entity directing diversity increase. Its originator is firmly opposed to such interpretations of the hypothesis.

bacterium (plural, **bacteria**) Prokaryotes that constitute a domain of life. Only a very small percentage are pathogenic to other organisms.

biodiversity The variety of life on Earth, in an ecosystem, or in a biological group. The biological variety and variability of life on Earth. It is a measure of variation at the genetic, population, species, and ecosystem level. It can be approximated as the number of species. One definition of it is the number of species and how equally individuals are distributed between species.

biological pump The sea's biologically-driven sequestration (burial) of carbon from the atmosphere and from land runoff, mainly via rivers, to the ocean interior and seafloor sediments.

biosphere The part of the Earth occupied by life; it is the sum of all Earth's ecosystems.

biota The animal and plant life of a particular region, habitat, or geological period.

bivalve mollusk A mollusk that has a shell of two parts that has a hinge, and that it can open and close; inside the shell is the soft-bodied animal. They are invertebrates. See mollusk.

calcium carbonate A chemical compound that contains calcium and carbon, is important in the carbon cycle and hence regulation of the amount of carbon in the biosphere, and is in echinoderms (the group that includes sea stars and sea urchins), coral, the shells of bivalve mollusks, the shells of some phytoplankton, and some other groups of organisms. Its chemical symbol is CaCO3. Limestone is mostly calcium carbonate and was mostly formed by biological organisms, sequestering great quantities of carbon.

Cambrian explosion   A geologically short interval of time approximately 541 mya, at the beginning of the Cambrian period (about 541 to 485.4 mya) when practically all major animal phyla first appeared suddenly and quickly.

carbon   The most important element for life. It is in all large biological molecules, including DNA, RNA, proteins, carbohydrates, and fats. It is also in carbon dioxide. Chemical symbol is C.

carbonate   A chemical compound that has carbon and oxygen. Many carbonates are used by life.

carbon cycle   The biogeochemical cycle by which carbon is exchanged between the biosphere, Earth, water, and atmosphere. It consists of a short-term and a long-term carbon cycle. See long-term carbon cycle, short-term carbon cycle.

carbon dioxide   A molecule used in photosynthesis and produced in cellular respiration. Animals, including humans, expel it when they breathe. It is the main and most abundant greenhouse gas, causing warming of the planet because people are adding too much of it to the atmosphere. People are also adding too much of it to freshwater ecosystems and the ocean, making both of these aquatic systems too acidic. Chemical symbol is CO2.

carbon sequestration   The removal of carbon from the biosphere and atmosphere and storing it elsewhere, such as in the Earth or under the seafloor. This includes weathering and burial. See weathering.

carbon sink   An entity that removes more carbon from the biosphere and atmosphere than it releases to them. In the case of the long-term carbon cycle, this carbon is stored in the solid Earth, including the seafloor.

carbon source An entity that adds carbon to the biosphere or atmosphere. This includes volcanoes, movement of the continents over the seafloor's calcium carbonate and organic sediments, and spreading of the seafloor. It also includes the burning of fossil fuels by humans.

catalyst (the verb is **to catalyze**) A substance that increases the rate of a chemical reaction, aids its occurrence, or both. An enzyme is a specific type of catalyst. See enzyme.

cell The basic structural and functional unit of life. All organisms except viruses are made up of cells.

cellular organism An organism made up of cells. All organisms except viruses are cellular organisms.

chloroplast The organelle of the cell that carries out photosynthesis in organisms that can photosynthesize, such as cyanobacteria, phytoplankton, and plants. Chloroplasts have their own genomes, hence their own DNA.

chromosome A structure found inside the nucleus of a eukaryotic cell, composed of most of the cell's DNA tightly coiled around proteins; it carries all of the genetic material of the cell, except for that in the mitochondria and chloroplasts. In prokaryotes, which do not have a nucleus, the DNA scattered throughout the cell can be called the chromosome.

class A taxonomic group of related orders of organisms. The taxon between order and phylum.

coevolution Evolution by positive feedback between two species, in which a change in the first species leads to a change in the second species, which leads to another change in the first species, and so on, until an equilibrium is reached in which there is no net gain in further changes in either species with respect to the traits undergoing coevolution. Coevolution may be positive or symbiotic, as in the evolution of pollination. Alternatively, it may be negative or antagonistic, as in predators and their prey, or two species competing with each other. There is also symbiotic genetic coevolution, which involves the transfer of genes between species. See symbiotic coevolution; antagonistic coevolution; symbiotic genetic coevolution.

commensalism (the adjective is **commensal**)   A relationship between two species in which one benefits and the other is not affected, being neither helped nor harmed. An example is a hermit crab using the shell of a dead snail. The hermit crab obtains a protective, moveable home, and the snail is not affected because it is dead.

control gene   See regulatory gene.

diversification   An increase in the number of species of a group (taxon) of organisms due to many speciation events. It often occurs when the taxon reaches a new adaptive zone with many new unfilled niches with no competition from other species, and no predators. It also occurs if the taxon invades a new habitat and thus has access to many new, unoccupied niches and a lack of competition from other species, and a lack of predators. Diversification due to these two reasons lasts only until all the new niches are filled and a new equilibrium with the environment is reached. Another term for diversification is adaptive radiation.

DNA   the genetic material of all organisms, except for some viruses. DNA codes directly for RNA and indirectly for proteins. In eukaryotes, the majority of it is in the nucleus of the cell, but it is also in the mitochondria and chloroplasts. Prokaryotes do not have a nucleus, mitochondria, or chloroplasts, so the DNA is spread throughout their cell, most of it on one circular chromosome, but some DNA can also be in smaller circular structures called plasmids in prokaryotes only. Some viruses have DNA as their genetic material; some have RNA.

domain   A group of related kingdoms. The highest taxonomic group. There are three domains: archaea, bacteria, and eukaryotes.

ecology   The study of the relationships of organisms and species to each other and to their environment. This includes humans. Ecology considers organisms at the individual, population, species, community, ecosystem, and biosphere levels. Relationships include predator-prey, parasite-host, competition between species and between individuals of the same species, commensalism, and symbiosis.

ecosystem A biological community of interacting organisms and their physical environment. An ecosystem consists of all the organisms and the physical environment with which they interact.

Ediacaran biota The life of the Ediacaran Period (about 635–541 mya). It was simple and all of it was in the sea. It included the first multicellular organisms. Many of the animals could not move, and were attached to rocks or the seafloor. Many of the animals were frond-shaped and tubular. They included trilobites and jellyfish. They were not as complex on the average as the animals that followed them, from 541 mya to the present.

endogenous retrovirus (ERV) A type of jumping DNA segment in the genome that resembles viruses and can be derived from them. ERVs sometimes get packaged and moved within the genome, with the result that they end up serving a vital role in gene expression and regulation for their host. They comprise up to 5–8% of the human genome. Apparently, some of them evolved from viruses, although some researchers think some viruses evolved from ERVs, since ERVs can mutate and become separate from their cell, and even pathogenic.

enzyme A catalyst that regulates the rate at which the chemical reactions of living organisms proceed, generally increasing their rate. Enzymes aid the chemical reactions of organisms. Without enzymes, many of these reactions would not take place at a perceptible rate. Enzymes regulate all aspects of cell metabolism. All enzymes are proteins. See catalyst.

et al. From the Latin phrase meaning "and others." A citation reading Jones, V.B. et al. means "Jones, V.B. and others".

eukaryote An organism whose cell or cells have a nucleus enclosed in a membrane. The nucleus contains the cell's DNA. Eukaryotes have organelles enclosed in membranes, such as mitochondria and chloroplasts.

Eukaryotes comprise one of the three domains of life, the other two being the bacteria and archaea, which are both prokaryotes.

eutrophication The depletion of oxygen in a freshwater or marine ecosystem that occurs because an excess of nutrients such as phosphorus or nitrogen are added, leading to a population explosion of phytoplankton, which causes a high population of bacteria that decompose the phytoplankton when they die in great numbers. The bacteria deplete the oxygen when they decompose the phytoplankton. The resulting system has very low oxygen levels, and so cannot support very much life. Eutrophication can lead to the death of many fish and invertebrates.

genome The complete set of genes or genetic material in a cell or organism.

evolution Change in the characteristics of a gene, genome, trait, population, or species over time. It is mostly the result of natural selection. It is how the great diversity of life came about. Life started as one-celled prokaryotes and evolved to produce viruses, trees, elephants, humans, and the rest of life.

exaptation The change in function of an already-existing gene, protein, structure, or trait due to natural selection. It may or may not involve changing its form. Evolution often works by taking existing genes, proteins, structures, and traits, and building on them, adapting them to new functions. An example of exaptation is the fins of fish used for swimming evolved into legs used for walking when amphibians evolved from fish.

exoskeleton The external skeleton of arthropods.

family A group of related genera. The taxon between genus and order.

feedback A process in which the effect alters its cause. The effect and cause may or may not have causes and effects between them. It makes a circular loop, called a feedback loop. Factor A affects factor B, which affects factor C, which affects factor A. See negative feedback, positive feedback.

food web The branching flow of energy in an ecosystem from primary producers, such as plants, to herbivores to carnivores to decomposers. See trophic level.

Gaia One of the Greek primordial deities, the personification of the Earth and ancestral mother of all life. The Gaia Hypothesis is named for this goddess.

Gaia Hypothesis The hypothesis of James Lovelock and Lynn Margulis that posits that life made the Earth, especially the atmosphere, favorable to life, and regulates it, keeping it favorable to life, by negative feedback. This hypothesis inspired and influenced the Autocatalytic Biodiversity Hypothesis/Pachamama Hypothesis.

genotype The complete set of genetic material, the entire genome, of an organism. It is all of the DNA of an organism (or all of the RNA in a virus whose genetic material is RNA). Genotype contrasts with phenotype (see phenotype).

genus (plural, **genera**) A group of related species. The taxon between species and family.

hectare A unit of area equal to 10,000 square meters, or 2.471 acres.

horizontal gene transfer (**HGT**), also called lateral gene transfer The lateral movement of genetic material between organisms other than from parent to offspring. It can occur between organisms of the same or different species, even between different domains and between viruses and cellular orgaisms. It can generate variability without disrupting the compatibility of the parts of the organism.

long-term carbon cycle The carbon cycle that involves the exchange of carbon between rocks and the Earth's surface; it operates over millions of years. See carbon cycle, short-term carbon cycle.

ibid. An abbreviation for the Latin word ib dem, meaning "in the same place", used in a scholarly reference to refer to the reference immediately cited beforehand. "It was shown (Smith, 2002) and it was also shown (ibid.)" means: "It was shown (Smith, 2002) and it was also shown (Smith, 2002)".

invertebrate An animal without a backbone. This includes insects, worms, sea urchins, and many other animals.

kingdom A group of related phyla. The taxon between phylum and domain.

marine snow A continuous shower of debris falling through the sea, originating in its upper layers, that resembles falling snow, and is made up of parts or all of dead or dying animals, seaweeds, plankton, bacteria, archaea, fecal matter, sand, soot, and other inorganic dust. Most of it is made up of and generated by organisms. Some of it gets buried in the seafloor, sequestering large amounts of carbon. It delivers nutrients that are used by organisms at the bottom of the sea.

Medea A sorceress in Greek mythology, who killed her children, so was a bad mother in these versions. Her name was used to name the Medea Hypothesis.

Medea Hypothesis The hypothesis of Peter Ward that proposes that life is suicidal and tends to hurt life. It disagrees with the Autocatalytic Biodiversity Hypothesis/Pachamama Hypothesis and the Gaia Hypothesis.

metabolism The set of life-sustaining chemical reactions in organisms, including obtaining energy from food, building molecules like proteins, and elimination of wastes.

meter The basic unit of length in the metric system, equal to just over three feet.

mineralization The process by which chemicals present in organic matter are decomposed or oxidized into easily available forms for plants. This is done by decomposers, including bacteria, soil mesofauna, soil macrofauna, and other groups.

mitochondrion (plural, **mitochondria**) The organelle of the cell that carries out cellular respiration, using oxygen to produce energy. Mitochondria have their own genomes. They are only in eukaryotes; prokaryotes do not have them.

mollusk (American spelling is mollusk, and British is mollusc.) An invertebrate of a large phylum which includes snails, slugs, mussels, clams, and octopuses. They have a soft, unsegmented body and live in aquatic or damp habitats, and most kinds have an external shell containing calcium carbonate.

morphology The physical form and structure of an organism. Feathers, wings, and colors are aspects of a bird's morphology.

natural selection The process by which organisms better adapted to their environment tend to survive with higher probability and produce more offspring than less well-adapted ones. It results from differences in the genotype and phenotype. It is a key mechanism of evolution. It can result in extraordinary adaptations. It is a major mechanism by which the Autocatalytic Biodiversity Hypothesis operates. It often causes an increase in biodiversity.

negative feedback A process in which an effect decreases the magnitude of its cause. It is stabilizing. In negative feedback, one factor, call it factor A, causes another factor B to increase, and factor B causes factor A to decrease. There can be more factors in the loop. So if A increases B, which increases C, which comes back and decreases A, there is negative feedback. Since it makes a circle, one often uses the term negative feedback loop to refer to the process. An example of negative feedback is: Predators eat more prey when there is more of it, and less prey when there is less of it, so predators stabilize their prey populations by negative feedback. See feedback, positive feedback.

neoteny   The sexual maturity of an animal while it is still in a larval state, or the retention of juvenile features in the adult animal. Larvae that underwent neoteny can reproduce. This can lead to macroevolutionary breakthroughs. The vertebrate line was started by neoteny in the sea squirt. The adult sea squirt cannot move and has no nervous system, but its swimming larva has a nervous system. This larva underwent a mutation that made it able to reproduce while still a larva; this started the vertebrate line.

niche, also called **ecological niche**   The role of a species in an ecosystem. For example, one aspect of the niche of the wolf is that it is a predator that eats deer, rabbits, and other animals.

nitrogen fixation   The process by which molecular nitrogen in the atmosphere (N2) is converted to a form of nitrogen such as ammonia (NH3) that can be used by plants. Nitrogen fixation can be carried out by lightning, but most is carried out by life, mostly by nitrogen-fixing bacteria, some of which live freely in the soil and some of which live in plant roots. The nitrogen-fixing bacteria are symbiotic with plants. The plant roots give them a home, and they provide the plant with nitrogen in a form that the plant can use.

nitrogen-fixing bacteria   Bacteria that fix nitrogen. See nitrogen fixation.

order   A group of related families. The taxon between family and class.

organelle   A specialized structure in a cell that carries out a specific function for the cell. It is analogous to an organ in the body. The mitochondrion is an organelle.

oxygenic photosynthesis   The chemical reaction that uses the energy of sunlight to convert carbon dioxide and water into oxygen and carbohydrates (food). Oxygenic photosynthesis is carried out mainly by green plants, phytoplankton, and cyanobacteria. It produces the carbohydrate that feeds a great deal of the biosphere. It is called *oxygenic* because also produces oxygen, the energy source for a tremendous number of species. See photosynthesis. Oxygenic photosynthesis is the opposite reaction of respiration.

Pachamama   The ever-present mother goddess of Earth who has her own self-sufficient and creative power to sustain life on Earth, revered by indigenous peoples of the Andes, such as the Quechua and Aymara. Pachamama is usually translated as Mother Earth, but a more literal translation is World Mother in Quechua and Aymara. The Pachamama hypothesis is named for this goddess.

Pachamama Hypothesis   See Autocatalytic Biodiversity Hypothesis

pathogen (the adjective is **pathogenic**)   An organism that causes disease.

phenotype   Characteristics and traits of an organism that can be seen. The term covers the organism's morphology, which is its physical form and structure. The phenotype also includes developmental processes, biochemical and physiological properties, and the organism's behavior. The phenotype includes the outward appearance (shape, structure, coloration, pattern, size, weight), and the form and structure of the internal parts, like bones and organs. Phenotype contrasts with genotype. See genotype.

photic zone   The top layer, nearest the surface of the ocean or a lake, where enough light penetrates the water to allow photosynthesis to occur. Photosynthetic phytoplankton and bacteria live in this zone.

photosynthesis   The chemical reaction that uses the energy of sunlight to convert carbon dioxide to carbohydrate (food). Oxygenic photosynthesis produces oxygen. Anoxygenic photosynthesis produces sulfur. See anoxygenic photosynthesis, oxygenic photosynthesis.

phylum (plural, **phyla**)   A group of related classes. The taxon between class and kingdom.

phytolith   A microscopic silica grain formed by plants in their leaves and stems. Plants that form them include all grasses, and some herbs, shrubs, and trees. A small amount of carbon becomes encapsulated in each grain. This sequesters carbon, effectively removing it from the atmosphere for millennia. Phytoliths are an effective mechanism by which life sequesters carbon.

phytoplankton Small, often unicellular, organisms in the sea and freshwater ecosystems that carry out photosynthesis. They are very important to the biosphere and the Autocatalytic Biodiversity Hypothesis because they are the base of aquatic food webs, produce much of the Earth's oxygen, and sequester a great deal of carbon. Their sequestering of carbon helps regulate Earth's temperature and the acidity of aquatic ecosystems.

placental mammal a mammal that carries the fetus in the uterus of the mother to a relatively late stage of development. Placental mammals contain the vast majority of living mammals. They are the most complex mammals, and include wolves, deer, lions, monkeys, and humans.

plasmids Small circular pieces of DNA in cells that are separate from the chromosome. They replicate independently. They contain genes, often ones that benefit the organism. They can transfer to other cells. They can integrate into the cell's chromosome (bacteria only have one chromosome). They are mostly in bacteria.

point mutation A mutation affecting only one base or a few bases that make up DNA.

positive feedback A process in which an effect increases the magnitude of its cause. It is destabilizing. Factor A causes factor B to increase and B causes A to increase. There can be more than two factors. If A increases B, which increases C, which loops back and increases A, there is positive feedback. Since it loops back into a circle, it is called a positive feedback loop. For example, people have caused increased CO2 in the air, warming the temperature, and causing the ice over the sea on the coast of Greenland to melt. This exposes the sea, which is darker than the ice that has melted and disappeared. Darker surfaces absorb more heat than lighter surfaces. So melting the ice and exposing the darker sea causes more heat to be absorbed, temperatures get hotter, and more ice melts. Positive feedback can lead to catastrophe. See feedback, negative feedback.

power of ten, or **powers of ten**   A shorthand way of writing a large number in which a number between 1 and 9 is multiplied by 10 raised to an exponent, or power. For example, 5 X 106 is 5 times 10 to the 6th power, which is 5 multiplied by 10 six times, or 5 with 6 zeroes after it, or 5,000,000. And 6.7 X 103 is 6.7 times 10 to 3rd power, which is 10 multiplied by 10 three times, or 6,700.

Precambrian supereon   Time in Earth's history spanning from just after the formation of the Earth about 4.6 bya to about 541 mya. Life began in this supereon. Complex animals appeared right after it ended.

predation   A ecological interaction where one species, the predator, kills and eats another species, its prey. It is a form of symbiosis if the population of the prey is considered, rather than individual prey organisms.

predator   An animal that eats other animals, its prey. Predators increase biodiversity. They regulate prey populations. They selectively eat better competitors among their prey species, preventing the species that is the better competitor from driving the poorer competitor locally extinct, thus maintaining high diversity. Removal of predators can result in a catastrophic trophic cascade with the collapse of the ecosystem. See trophic cascade.

prey   An animal that is eaten by another animal, its predator.

prokaryote   A unicellular organism whose cell lacks a nucleus, having its DNA existing throughout its cell. Prokaryotes have no organelles, such as mitochondria. They are generally unicellular, and consist of two domains, the bacteria and archaea.

protein   Large biological molecules made up of amino acids that are used by organisms either for structure or to run chemical reactions. Proteins that run chemical reactions are called enzymes. Proteins are major molecules of life and necessary for life to exist. The order of amino acids in proteins determines their nature and function, and is coded by DNA and RNA. RNA codes for proteins directly. Since DNA codes for RNA, DNA codes for proteins indirectly.

protist A member of a large group of eukaryotes that are mainly unicellular and include amoebas, diatoms, and slime molds. They are not a natural group; rather, the biological category protist is used for convenience.

protozoa (singular is **protozoan**) An informal term for a group of unicellular eukaryotes. They are not a true taxon. They include the *Amoeba*, *Paramecium*, and *Euglena*.

pseudogene A DNA sequence that resembles a gene and was formerly a gene, but has been mutated into an inactive form with no function. Pseudogenes that are paired with a functioning gene that they were duplicated from can evolve freely without harming the organism they are in, and can potentially mutate into a gene with a new adaptive function.

recombination, also called sexual recombination The exchange or trading of segments of DNA between chromosomes in sexual reproduction. It can increase DNA content and genetic variability while maintaining the compatibility of the different parts of the genome. The author of this book has proposed that this is the principal reason why sexual reproduction evolved and is usually favored by natural selection.

reduced carbon Carbon that is not combined with oxygen. The carbon in carbon dioxide, CO_2, is combined with oxygen, so it is not reduced carbon. The carbon in methane, CH_4, is not combined with oxygen, so is reduced carbon. Sequestration of reduced carbon increases atmospheric oxygen levels because it decreases the amount of carbon that can combine with and hence deplete the free molecular oxygen in the atmosphere, and it does so without any loss of oxygen from the atmosphere. (It also lowers the temperature). Sequestration of carbon that is not reduced lowers the temperature, but does not increase atmospheric oxygen levels because it buries oxygen with the carbon.

regulatory gene A gene that regulates one or more other genes, turning them on or off.

respiration, also called **cellular respiration** A reaction that converts oxygen and carbohydrate to carbon dioxide and water, producing energy. In eukaryotic cells, it takes place in the mitochondrion. It produces the energy needed for complex organisms to exist and thrive. Respiration is the opposite reaction to oxygenic photosynthesis. (Respiration can mean breathing, but is not used in that context in this book.)

RNA Biological molecule that acts as an enzyme or, in some viruses, as the genetic material. Normally, DNA codes for it, and it codes for proteins. Sometimes, it can code for DNA. Some viruses have RNA as their genetic material; some have DNA. It is thought that RNA was the first molecule in the chemical evolution of life, since it can act as both genetic material and an enzyme.

sequence, also called **DNA sequence** Noun: The order of the four chemical bases in a DNA molecule of an organism. The bases are abbreviated A, C, G, and T. Verb: To determine the order of the four chemical bases in the DNA of an organism by the use of technology and chemistry.

sexual reproduction The production of new organisms by the combination of genetic information of two separate organisms. It can generate variability while maintaining the compatibility of the parts of the genotype and phenotype.

short-term carbon cycle The carbon cycle involving the short-term carbon reservoir, in which carbon is stored in the atmosphere, oceans, and biosphere, with the ocean containing the largest amount of carbon. It takes months to centuries to recycle carbon through the short-term reservoir. See carbon cycle, long-term carbon cycle.

species A group of organisms with similar characteristics. A group of organisms that can breed with each other and produce fertile offspring in nature.

structural gene   A gene that does not regulate other genes.

subspecies   A taxonomic category or rank below species, used for populations that live in different areas and differ in physical characteristics, but are the same species.

symbiosis   A relationship between two (or more) species in which each benefits the other. An example is pollination, where an animal such as a bee obtains pollen and nectar for a nutritious meal from the flower and the plant is pollinated by the bee and so can reproduce.

symbiotic coevolution   Term coined by the author of this book, meaning evolution between two species involving feedback and that is beneficial to both species. Pollination of plants by animals and dispersal of seeds by animals both came about by symbiotic coevolution.

symbiotic genetic coevolution   Term coined by the author of this book, meaning coevolution between two symbiotic species in which at least one of the species has its variability enhanced by the transfer of genes from one species to the other.

taxon (plural, **taxa**)   A group of organisms used for classification of any rank, such as a species, family, class, phylum, or domain. Variety, subspecies, and species are the lowest ranks, and domain is the highest rank.

taxonomic   Concerning the classification of biological organisms.

transcription factor   A protein that regulates the production of RNA by DNA, by binding to DNA, usually enhancing this RNA production.

transduction   A virus can integrate its DNA into the DNA of its host. Transduction is the process by which such integrated viral DNA takes some of its host's DNA with it when it leaves its host's DNA and brings it with it when it integrates its DNA into a new host. This adds DNA to its new host. This can lead to new adaptive evolutionary changes in the host that receives the DNA. Transduction is usually between host of the same species, but can be between hosts of different species.

trophic This refers to the feeding and nutrition relationships of different species in a food web.

trophic cascade A catastrophic loss of species in an ecosystem as a result of the removal a key species in the food web, such as a key predator like the gray wolf (*Canis lupus*), or a key primary producer such as a species of phytoplankton.

trophic level The position of a species in a food web. A food web starts at trophic level 1 with primary producers such as plants that obtain their energy from sunlight via photosynthesis and nutrients from the soil. Level 2 is herbivores that eat plants, level 3 is carnivores that eat herbivores, and level 4 is carnivores that eat other carnivores and perhaps herbivores. The last level consists of the decomposers, such as vultures, fungi, and bacteria, which recycle the nutrients back into the soil, making them available to plants again. Ecological communities with higher biodiversity form more complex and branched trophic paths and food webs. See food web.

vertebrate An animal with a backbone. This includes fish, amphibians, reptiles, birds, and mammals.

virulence The harmfulness of a pathogenic organism, such as a disease-causing virus, bacterium, or fungus.

virus A submicroscopic infectious agent that replicates only inside the living cells of an organism. Viruses cannot live on their own. Viruses infect all life forms. They have a capsid of protein surrounding their genetic material. Their genetic material can be either DNA or RNA. They are the most diverse life forms, and have greatly enhanced the evolution and increased the diversity of cellular life forms. There is controversy over whether they are living organisms. I define life as that which can undergo natural selection, so consider viruses to be living organisms.

viral shunt   The constant attacking and killing of prokaryotes by viruses in the sea. It recycles nutrients and greatly fertilizes the sea. Marine phages kill 20 to 40% of ocean bacteria every day

weathering   The breaking down of rocks, soil, minerals, wood, or even artificial materials through contact with the Earth's atmosphere, water, and biological organisms. In it, CO_2 is combined with minerals in chemical reactions. The resulting material is then carried via rivers and creeks to the sea, where it falls to the ocean bottom and is buried. Weathering is a key step in sequestering carbon, removing it from the biosphere. Weathering is the main way carbon is removed from the atmosphere nonbiologically. Weathering can be done biologically too, which also sequesters carbon.

ABOUT THE AUTHOR

David Seaborg is an evolutionary biologist. His undergraduate degree is from the University of California at Davis in zoology, and his graduate degree is from the University of California at Berkeley, also in zoology. He is a world-leading authority on biodiversity. He originated the concept that organisms act as feedback systems in their evolution, and that they thus play an important role in their evolution. This concept is a mechanism for how species can remain unchanging for long periods of time and also undergo rapid, enormous evolutionary change. He showed that the canonical genetic code is not optimal for evolution, and is stuck on an adaptive peak, and how populations cross over maladaptive valleys from one adaptive peak to another. He published a hypothesis to explain how homosexuality evolved even though it tends to reduce the number of offspring produced.

He wrote two books for scientists and educated laypeople on his original, innovative, radical hypothesis that species shape their environment to make it better for life under natural conditions, and that organisms formed the high-oxygen atmosphere, the ozone layer, the soil, and regulate Earth's temperature, making the evolution of higher life possible. Humans are the exception to this hypothesis. The books are *How Life Increases Biodiversity: An Autocatalytic Hypothesis* and *Organisms Amplify Diversity: An Autocatalytic Hypothesis*, both published by CRC Press/Taylor and Francis Group (available at their website and on Amazon books).

He has taught biology at all levels from kindergarten to the university level, including at the University of California at Berkeley.

David is an environmental leader. He founded and is President of the World Rainforest Fund (worldrainforest.org), a nonprofit, tax exempt foundation dedicated to saving the Earth's tropical rainforests and biodiversity. This organization saves rainforest by empowering the indigenous people of the rainforest. Scientific studies have shown that this is the most cost-effective and long-lasting way to save rainforest. The World Rainforest Fund has saved rainforest in the Amazon River system of Brazil, Columbia, Ecuador, and Peru, and in the Democratic Republic of the Congo, the

Philippines, Malaysia, and Borneo. It set the record for the most species saved per dollar when it helped stop a road that if built would have resulted in the destruction of a 10,000-acre rainforest in Ecuador that has the highest biodiversity of any ecosystem on Earth according to scientists at the Missouri Botanical Garden. Had the road been built, exploiters would have used it to access the rainforest and destroy it. The World Rainforest Fund spent only $3,500.00 to stop the road and save this rainforest.

He raised funds to successfully help save Acalanes Ridge, a pristine hillside oak and grassland habitat in Lafayette, California.

He wrote an article that is a summary of the scientific research on the effects of high atmospheric levels of carbon dioxide other than global warming. Unlike the climatic effects, these effects are not well known to the general public. They are very serious, and have the potential to cause high levels of extinction of species and greatly disrupt ecosystems and our food supply.

He was on the city of Lafayette's General Plan Advisory Committee, where he played a key role in producing a ten-year General Plan for that city that emphasized environmental sustainability, preserving open space, combating global warming, and energy conservation.

In the 1990's and part of the first decade of this century, he served on the Board of Directors and as Vice President of the Club of Rome of the USA, the environmental think tank that published the Limits to Growth in the 1970's. This is a computer simulation study that showed that continued growth and consumption of resources will lead society to disaster. He is currently on the nominating committee for the Goldman Environmental Prize, the most prestigious grassroots environmental prize in the world.

He conceived, and helped secure passage by the Berkeley City Council, an ordinance banning the use of old growth rainforest and redwood in all products used by the city of Berkeley. This ordinance also required all businesses contracting with Berkeley to stop using old growth rainforest and redwood in any products or services Berkeley hires them to use or perform, or in any product they sell to this city.

David carried the Ten Commandments for the Earth, a version of the original Ten Commandments re-written to focus on saving the Earth's environment, while riding a camel down Mount Sinai, the mountain in Egypt down which Moses allegedly carried the original Ten Commandments. Then, in a brief ceremony, he presented these Ten Commandments to a Bedouin youth, who represented the indigenous people and the youth of the planet, the generation inheriting the Earth for its stewardship.

David conceived the idea for and was the head organizer for a press conference of Nobel Prize winners on global environmental and poverty issues that was held at the time of the 100th Nobel Prize ceremonies in Stockholm, Sweden, in December, 2001.

He is an award-winning nature and wildlife photographer and an award-winning poet. He wrote a popular and acclaimed poetry book called *Honor Thy Sow Bug*. He is listed is in *Who's Who in America*. An excellent public speaker, he lectures to various scientific, environmental, civic, business, and other organizations on evolutionary biology, the philosophical implications of science, and environmental issues.